Tractor Operator

Tractor Operator

Aakash Kumar
Priyanka M.

2026
Agri Horti Press
New Delhi-110 084

ISBN: 978-93-49690-88-2 (Hardbound)

Published By

Agri Horti Press

New Delhi

E-mail: agrihortipress@gmail.com

Preface

Tractor Operator is a comprehensive guide aimed at providing a detailed understanding of the fundamental principles, operational techniques, safety protocols, and maintenance practices related to tractor operation. This book is designed for both novice and experienced tractor operators, offering a blend of theoretical knowledge and practical insights to enhance the skills and safety of those working with tractors in agricultural and industrial settings.

The text begins with an introduction to the essential components of a tractor, covering the basic structure, engine types, transmission systems, and implements commonly used in conjunction with tractors. It delves into the different tractor models available in the market, highlighting the features and specifications that distinguish them from one another. The book also provides an overview of the types of attachments and implements, such as plows, harrows, seeders, and mowers, that are typically used in agricultural operations, ensuring that operators understand the tools at their disposal and how to optimize their use for various tasks.

A central focus of *Tractor Operator* is the practical aspect of tractor operation. The book provides step-by-step instructions on how to operate a tractor safely and efficiently, emphasizing the importance of proper technique in handling the machinery. It covers key areas such as starting and stopping the engine, controlling speed, shifting gears, steering, and using the hydraulics system for attachment control. The book is structured to offer clear, actionable guidance on how to perform common agricultural tasks, from plowing fields to transporting goods, with detailed explanations on how to ensure each task is done effectively and safely.

Safety is a critical aspect of tractor operation, and the book dedicates significant attention to ensuring that operators understand the potential risks associated with operating heavy machinery. The text provides a thorough overview of tractor safety measures, including the proper use of seat belts, rollover protection structures (ROPS), and the importance of wearing appropriate personal protective equipment (PPE). It also discusses common hazards such as uneven terrain, operating on slopes, and avoiding entanglement with moving parts. Emphasizing proactive safety measures, the book ensures that tractor operators are well-informed about the necessary precautions to minimize risks and accidents.

In addition to safety, *Tractor Operator* highlights the importance of routine maintenance and troubleshooting. The book offers a detailed guide on the maintenance tasks essential for keeping a tractor in good working condition, including checking oil levels, cleaning air filters, inspecting tires, and ensuring that the hydraulic systems are functioning correctly. It also provides troubleshooting advice for common mechanical issues, equipping operators with the knowledge to address minor problems before they escalate, thus ensuring the longevity and reliability of the tractor.

The book further explores the economic and environmental impact of tractor operations. It discusses how efficient tractor operation can contribute to greater productivity and cost savings in agricultural operations. The text emphasizes the importance of fuel efficiency, proper handling of machinery to reduce wear and tear, and effective management of resources to enhance the sustainability of farming practices. It also touches on modern advancements in tractor technology, such as GPS-guided tractors, automated driving systems, and electric-powered tractors, providing insight into the future of tractor operation and how emerging technologies may revolutionize the industry.

Throughout the book, a variety of real-world examples, case studies, and illustrations are included to help readers visualize key concepts and apply their knowledge to practical situations. Whether it is navigating challenging terrain or operating a tractor in varying weather conditions, these examples provide valuable insights into the complexities of tractor operation.

In conclusion, *Tractor Operator* serves as a comprehensive resource for anyone seeking to gain a deeper understanding of tractor operation. From safety measures to operational techniques and maintenance practices, this book equips readers with the knowledge and skills needed to become proficient and responsible tractor operators. It is an invaluable tool for those working in agriculture, construction, or any field where tractor use is integral to daily operations. Whether you are a beginner learning the basics or an experienced operator looking to refine your skills, this book will serve as a practical guide to mastering the art of tractor operation.

Aakash Kumar

Priyanka M.

Contents

Unit 1: Operation of the Tractor

Chapter 1

Daily Checking

1.1 Lubricate rear axle bearing (daily/periodically as per the equipment manufacturer's recommendation)

1.1.1 Lubricating Bearings

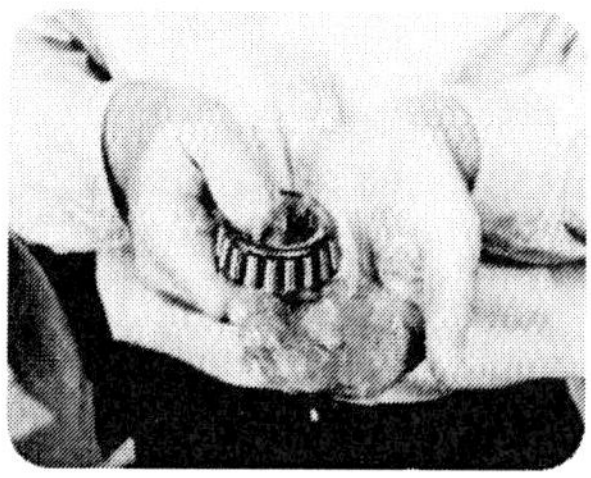

- Clean the bearing with solution and dry it.
- Properly see the bearings for wear and tear.
- Now dip the cleaned bearings in a protective lubricant or apply lubricant on the bearing using the brush.
- Rotate each bearing so that grease on it should be properly spread.
- After applying lubricants, wrap them with waterproof paper and place in a clean box.

Some tractors have rear axle bearings which are lubricated from the fluid in the transmission case.

Tractor rear axial bearings should be lubricated after every 500 hours as required or after every 10 hours, when the tractor is used on wet or muddy soil.

1.1.2 For greasing the real axle bearing:

❖ Open the tyre of the tractor by unscrewing the nut.

❖ Remove the tyre

❖ You will find the bush and the bearing

Apply grease or lubricating oil at the surface of bush and bearing

Check tractor rear axial bearing and shaft and apply grease, lubricants or engine oil to all its parts and then fit the tyre.

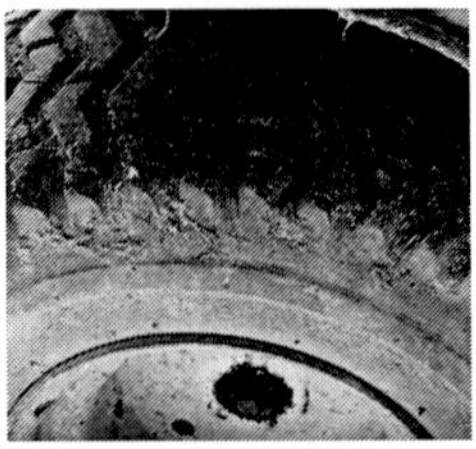

1.2 Check engine oil level

1.2.1 Engine Oil level

Engine oil level can be checked by inserting an oil dipstick in the engine and then pulling it up, which shows the oil level as indicated on the stick.

In the dip stick, high and low marks are present, which shows an exact oil level.

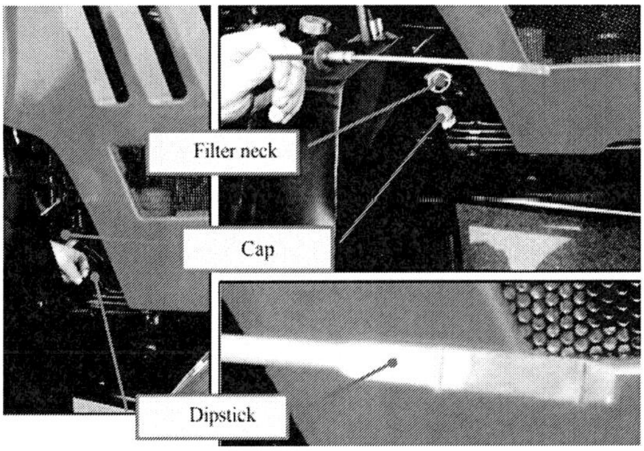

Fig Checking engine oil level

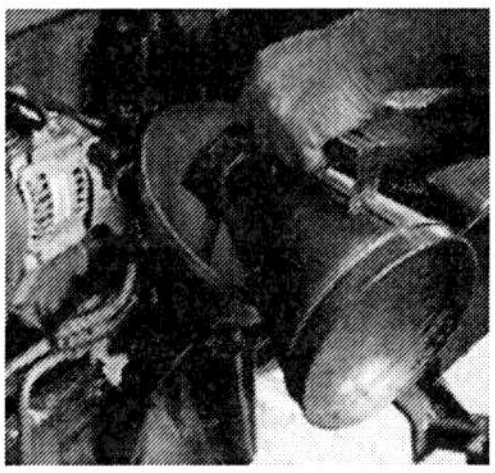

1.2.2 For checking engine oil level:

- ❖ Pull the dip stick located on the right side of the engine and clean it using a cloth.
- ❖ Insert it in engine oil tank and then pull out to see the oil level between upper and lower marks.
- ❖ If the level of the engine oil is low, add the new oil till required level marked.
- ❖ If you have top up the engine oil, then again check the oil level by inserting the dip stick.
- ❖ Always check the engine oil level before the start of engine for work.

1.3 Check transmission oil level

Tractor transmission oil should be checked at regular interval after several hours of operation to avoid any danger or damage. Checking of transmission oil should be done after an hour of working, so that the oil settles before checking the level in order to have accurate oil level reading on dipstick.

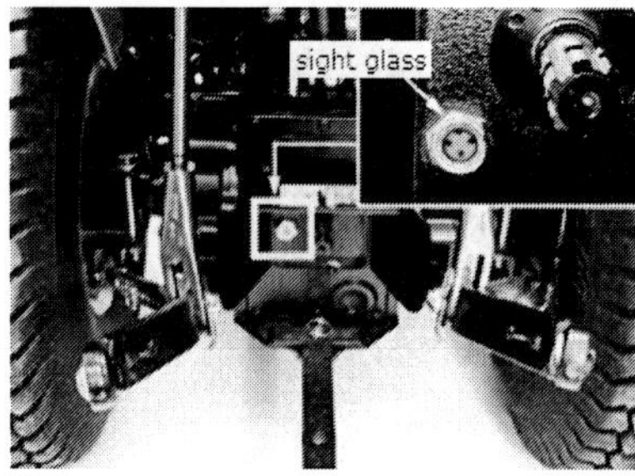

1.3.1 For checking oil level:

- Place the tractor on safe place.
- Allow the machine to cool down for at least one hour.
- See the sight glass on rear of transmission next to the rear PTO stub shaft.
- Oil should be maintained to a mid-level range on the sight glass.
- Refill the transmission with clean oil.
- Pull out the dipstick and clean it with a fresh cloth.
- Replace the dipstick by pushing it fully in and check that the oil should reach the upper mark on the dipstick when it is removed from the transmission aperture again.

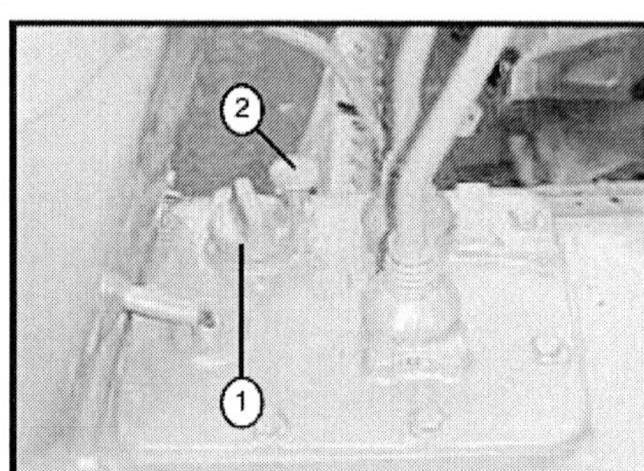

1. Breather cumfiller 2. Dipstick

- After refilling the transmission, start the engine to allow the oil to circulate throughout the transmission.
- After a minute, stop the engine and recheck the oil level by means of the dipstick.
- If oil level is low, top-up the oil till the upper mark of the dipstick.

1.4 Check steering box oil level

1.4.1 Steering box

Steering-box system is normally mounted in engine bay which is bolted to the side of the bodywork or to the front cross member.

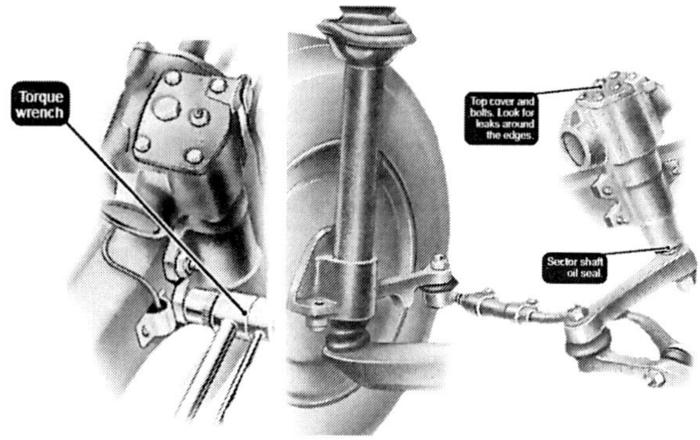

Checking steering box involves:

- ❖ Raise the front of tractor by keeping its weight on the wheels which can be done by driving the tractor on the ramps and carefully inspect the pit.
- ❖ Checking the steering box stiffness which happens due to oil leakage.

1.5 Check coolant level

Coolant is used in the radiator to cool the engine. When the tractor becomes heated up, in such case, you need to check for the radiator. Hot tractor radiator can be cooled down by water.

Check the coolant level in the radiator, once the radiator gets cooled.

1.5.1 For checking the coolant level:

❖ Place the tractor on level surface and apply brakes.

Open the tractor hood to see the radiator.

❖ Remove right side engine shield from tractor.

❖ Now check the coolant level in the radiator.

❖ In radiator, coolant level is marked with full line (A) and low line (B) on coolant overflow tank.

❖ Check the coolant when tractor engine is cold, as in such case, coolant level will be at low line (B) placed on coolant overflow tank.

❖ Also check the coolant in coolant box by remove the cap (C).

❖ If the coolant in the box is low, add the required ratio of coolant along with some amount of water.

❖ Close and tighten the overflow tank cap.

1.6 Drain water and sediment from fuel filter

Fuel filter has fluid level sensor which keeps track of the amount of water that has been collected.

1.6.1 To check the fuel filter of tractor:

❖ Place the tractor in safe place.

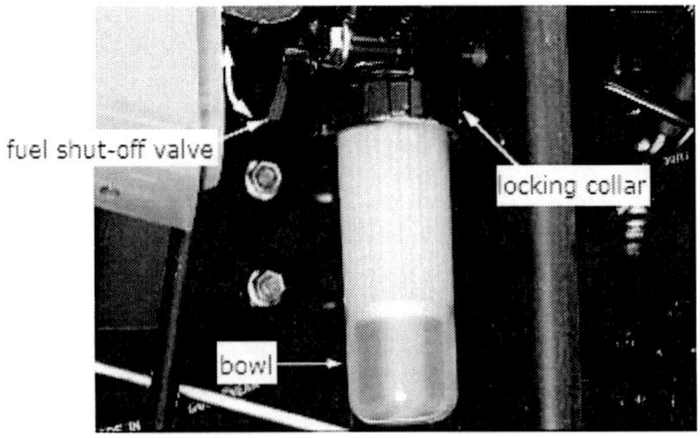

Allow the engine to get cool down.

- ❖ Raise the bonnet.
- ❖ Remove the side panel from right side of an engine.
- ❖ Close fuel shut-off valve.
- ❖ Keep the drain pan under the fuel filter sediment bowl to remove the fuel spillage.
- ❖ Move the locking collar counter clockwise in order to remove the bowl.
- ❖ Now remove and discard the fuel filter.
- ❖ Remove and keep the plastic ring and spring from the sediment bowl.
- ❖ Clean the bowl, plastic ring and spring.
- ❖ Place the plastic ring and spring back in original position in sediment bowl.
- ❖ Place new filter to the filter head.
- ❖ Keep the sediment bowl and locking collar in the position.
- ❖ Tighten locking collar to filter head to secure.
- ❖ Now open the fuel shut-off valve to install the side panel.
- ❖ Make the bonnet closed.

1.7 Lubricate tie rod ends

1.7.1 Tie rods

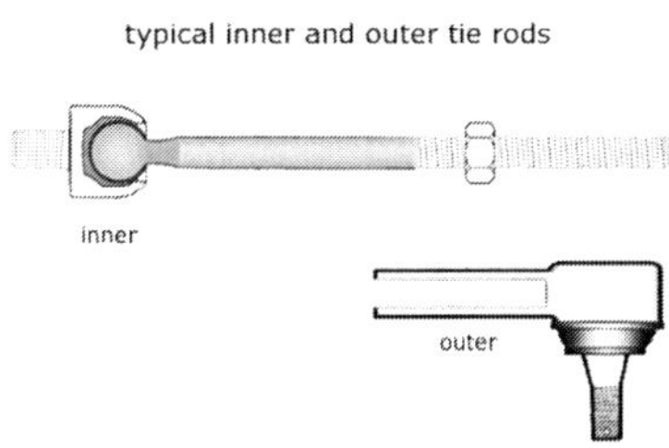

Tie rods are similar to small ball joints which allows up and down movement and rotation of the steering components.

- ❖ A tie rod can both turn and pivot which maintains the connection between steering and wheels as the suspension moves.
- ❖ Tie rods normally come in pairs, an inner and an outer.
- ❖ The inner tie rod will be closer to the center of the vehicle while outer tie rod is nearer the wheel.
- ❖ Threads are used for connecting the inner and outer tie rods.

1.7.2 Lubrication

- ❖ Lubricate tractor with grease fittings after every 50 hours of operation
- ❖ Lubrication of tractor grease fittings should be done after every 10 hours of operation or on daily basis when used in mud.

- ❖ Tie rods holds the lubricant in and debris out with flexible rubber boot around the tie rod base.
- ❖ Too much grease also causes the protective boot to tear.

- ❖ Lubricate the Left Tie Rod End and Right Tie Rod End with grease or engine oil.

1.8 Lubricate Steering spindles

The steering spindles are the part of the front axle on which the front wheels revolve. The steering spindles are sometimes fitted with either roller or ball bearings.

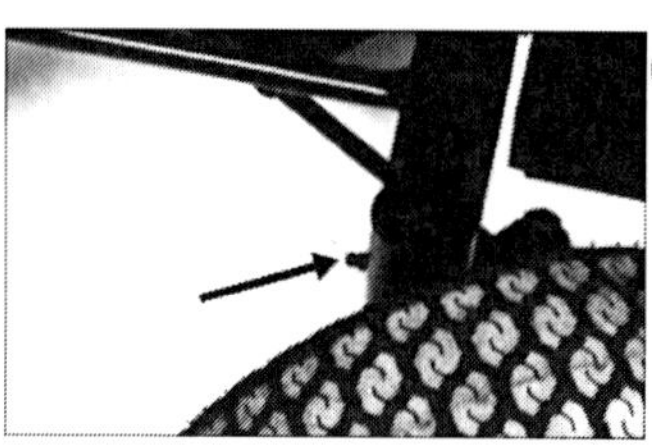

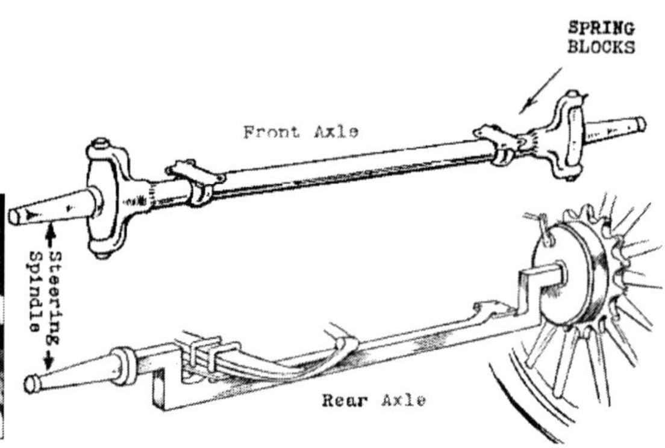

1,8,1 Lubrication

For lubricating tractor steering spindles:

- ❖ Place the tractor on leveled surface.
- ❖ Lift the front part of it using the jack.

Locate the steering spindle fitting on front of rear and front axle.

- ❖ Lubricate the spindle linkage using standard grease or lubricating oil.
- ❖ Also lubricate steering spindle fitting on other side of the tractor axle.
- ❖ Do lubricate on the left and right rear axle spindle.
- ❖ Release the jack.

1.9 Lubricate front axle pivot pin

Keep the tractor on level surface and use jack to lift the front part of the tractor.

Unscrew the nut

Remove the front axle out from the tractor

In the front axle, unscrew the nut

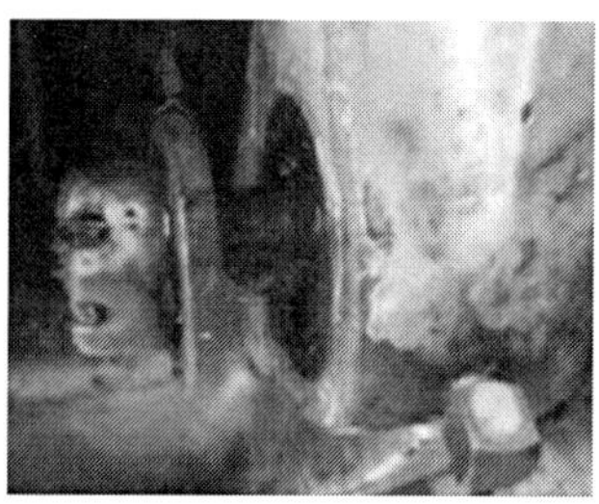

Lubricate the front axle pin with suitable grease or lubricating oil.

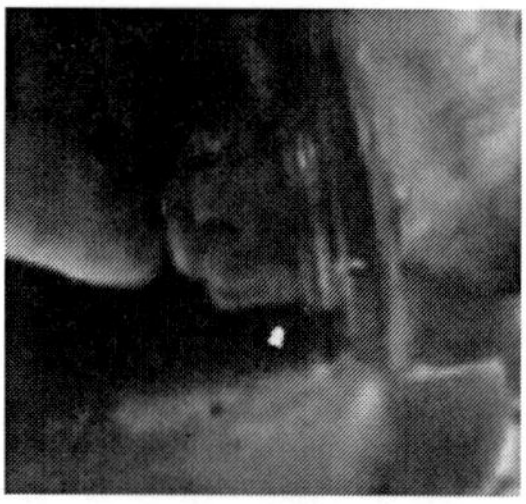

PC.10 Lubricate clutch linkage, throttle linkage hand, foot linkage and all hinges

1.10 Clutch Linkage

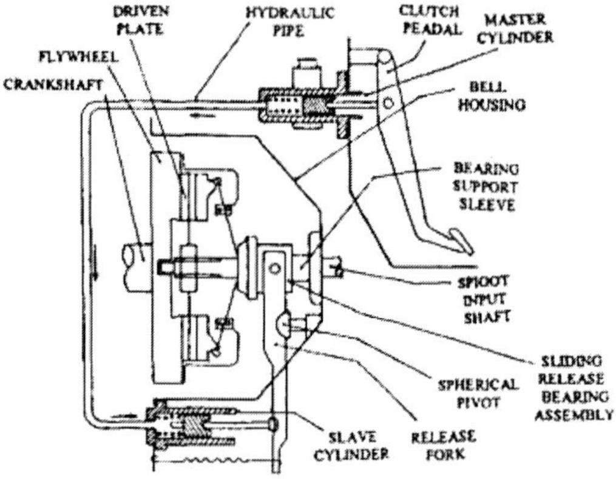

Clutch linkage in tractor is an arrangement of mechanical and hydraulic components. In tractor, the mechanical clutch linkage consists of:

- clutch pedal,
- series of linkage rods and arms
- cable

In this, hydraulic clutch linkage includes:

- clutch master cylinder
- reservoir
- hydraulic line
- slave cylinder

10.1.1 Lubricate clutch linkage

Tractor operator should refer to owner's manual for maintenance and lubrication of clutch linkage. Lubrication of clutch linkage parts should be done by tractor operator after every 50 hrs of tractor running. Proper lubrication of clutch linkage will help the tractor operator to solve many problems related to:

- stiff clutch pedal
- clutch slippage
- premature wear of throw-out bearing
- gear clash on shifting
- hard release of clutch

Tractor operator should carefully apply grease of standard quality and standard as prescribed in owner's manual to all the functional parts of the clutch linkage, so as to remove any such problems that exists from time to time under periodic maintenance.

1.11 Clean pre-cleaner

In tractor, pre-cleaner remove contaminant of different sizes from entering into the intake duct which can be done without using engine power. The purpose of the pre-cleaner is to:

- extends air filter life
- reduces air filter servicing and replacement
- lowers cost per operating hour
- separates more than 99% of 20 micron and above particles

The working of the pre-cleaner is that:

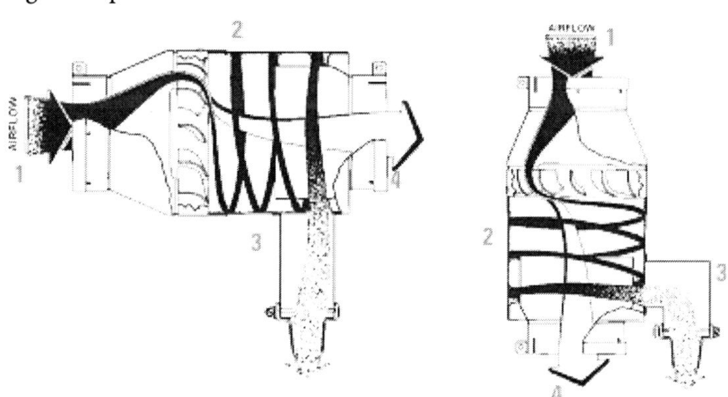

- When moisture and/or dust-filled air enters at one end, the built-in, stationary vanes cause the air to spin.
- Such spin creates centrifugal force, which pushes all moisture and dust to the outside wall where it separates from the air.
- Moisture and dust are thrown into the vacuator valve tubing, which automatically releases by the vacuator valve.
- Clean air (acceptable for maximum filter life and engine performance) passes to the air cleaner.

1.11.1 Cleaning

Pre-cleaner unit should be cleaned with either water, mild-soapy water or compressed air. Tapping or hitting the pre-cleaner components to dislodge contaminant may result in damage and can create difficulty in reassembling. The pre-cleaner/rain cap should be cleaned over the time which can be done by:

- ❖ Turning the engine off.
- ❖ Loosening of connecting clamps (metal pipe and scavenge hose) and then removing the cap of pre-cleaner. Cover or plug the intake pipe in order to save the air intake system from contamination during servicing.
- ❖ Turn the pre-cleaning unit upside down and remove the screws and disassemble the unit.

- ❖ Clean all the parts to remove dust and debris from each component.
- ❖ After cleaning, check the gasket on the perimeter of the upper baffle. If you find the damage in any of the parts, replace it. Also, check the gasket position and make sure it is installed evenly around upper baffle perimeter.
- ❖ After doing this, reassemble the components.
- ❖ With all components together, reinstall and torque the 6 screws to 2.3 ± 0.3-0.6 Nm. While doing this, use the removable screw adhesive on the screws if original blue patch has been worn off.
- ❖ Replace the cap on intake stack, and reconnect scavenge hose.
- ❖ Tighten the clamps to torque specifications. See if scavenge support gets disconnected, then reconnect it again.

1.12. Check for leakages if any

Before working on next day with your tractor, it is the duty of tractor operator to quickly examine for any leakage which can be in terms of fuel, oil, coolant, air pressure etc. Pre-operational of tractor involves checking of leakages or loosening of parts. By checking your tractor before using it, you could prevent costly repairs, down time and aggravation.

1.12.1 Types of leakages

- **Fluid leaks**: Look for any fluid leaks on the ground beneath the tractor. Leakage of liquid could be for coolant, engine oil and hydraulic oil levels.
- **Hydraulic system leak**: Look for leakage in hydraulic systems such as hydraulic joints, rams, pumps and hoses.
- **Gas leaks**: In lawn tractors, gas leakage appears in air filter which shows that engine valves are not working properly. If the tractor engine has fuel shut-off valve, make sure that this valve is closed when the engine is not in use. Check for any signs of valve damage which could prevent it from automatically closing when the engine stops running. If it stays open and sticks, it could indicate that the valve is worn-out and in need of a replacement.
- **Radiator leaks**: Check for radiator leakage which can be due to hole in the radiator, loose nuts, cracked body, overheating. If the radiator leaks, the coolant what is inside it, tends to leak drop by drop.
- **Engine block core plugs leaks**: Check for leakage in sides of engine blocks which results due to rusty streaks.

1.12.2 Checking the leakage

- For checking the leakages, turn off the engine.
- Clean the leakage area with cloth.
- Restart the engine so that the hydraulic system functions.
- Inspect for the leakage such as drips or trickles of oil or fluid.
- If the leak is still not visible, again clean the suspected areas.
- Check for leakages while running an engine.

1.13. Check the working of all lights

In a tractor, there are different lights for different purposes, Tractor operator should spent few minutes to check all the working of lights by making them ON or by applying pressure. It is important to keep the routine of checking of lights on day to day basis, as it will help in getting idea, which light is working and which needs a replacement.

Lights on a tractor are grouped as:

- driving lights
- work lights

1.13.1 Headlights

Headlight is an important driving light which is normally used while driving a tractor. The light can be operated or checked by using a switch that has two positions OFF and ON. While making your light switch ON, the current will pass through the light circuit as shown.

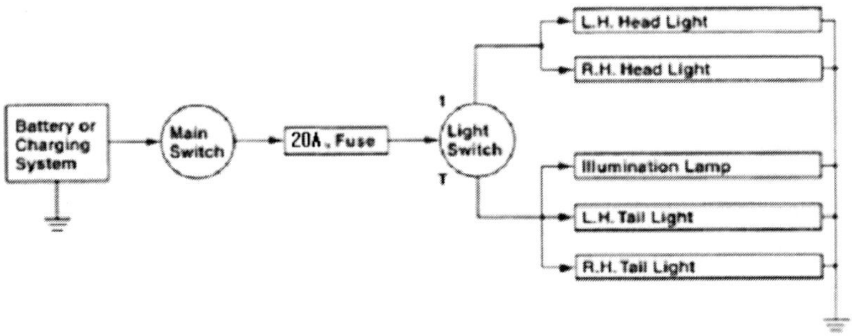

For checking the headlight of a tractor, an operator has to make sure that:

- ❖ Centers of the headlight low beams should point on the road atleast 30 meters ahead of the tractor,
- ❖ The left angle of the light should be straight ahead while the right angled should ahead slightly toward the edge of the road.
- ❖ Headlight high beams should aim just below the horizon.

Adjusting the headlights

- ❖ Before adjusting the headlight, check that the tires are inflated properly and that the tractor is carrying usual load.
- ❖ Park the tractor on level surface such that the headlights should be 25 centimeters from a wall or garage.
- ❖ Turn on the low beams and draw a cross on the wall to mark the center of each beam.
- ❖ Now back the tractor perfectly straight 7.5 meters.
- ❖ Each headlight has two adjusting screws, one at the top or bottom and one at the side to adjust the aim, tighten or loosen the screws.
- ❖ Now set each headlight at low beam to 5 centimeters below the horizontal lines with left even with vertical line while right 5 centimeters to right of vertical line.
- ❖ On four light system, adjust the high beams in same way as low beams.
- ❖ On two light system, high beams are automatically adjusted with low beams.

1.13.2 Roof light

Many tractors have roof light for better vision. The switch of the roof lights is present on the tractor dashboard.

Checking and Adjusting Roof light:

- ❖ In vertical direction, a point of illuminated area in plane of road towards left from longitudinal vertical plane passing through the headlight centre should not be more than 30m from the front outline of the tractor.

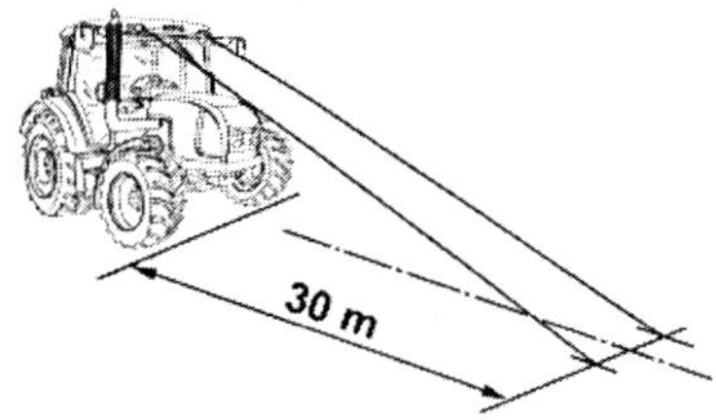

- ❖ In horizontal direction the headlight beams shall be parallel with the longitudinal axis of symmetry of the tractor.
- ❖ Carry out checks of adjustment of the headlamps at un-laden weight of the tractor.
- ❖ The front roof headlights may be used during traffic on ground roads only if there is a front-carried mechanism is coupled to the tractor or an equipment covering the main headlights.

1.13.3 Brake lights

Brake lights depend on tractor's electrical system which runs on series of switches and fuses to protect circuits. It works, when an operator press the tractor brakes. In tractors, mostly there are two brake lights behind and at the corners of the tractor body.

- ❖ If one or more of tractor brake lights isn't working properly, in such case:
  - brake light system fuse is blown
  - brake light bulbs are burned out
  - brake light wiring switch is broken

1.13.4 Back Light

Back light is an important light in the tractor as it gives the judgment to the tractor operator while he/she backs the tractor. Tractor operator should regularly check the back lights and in case, these lights are not working, he/she should immediately replace it.

1.13.5 Dashboard light

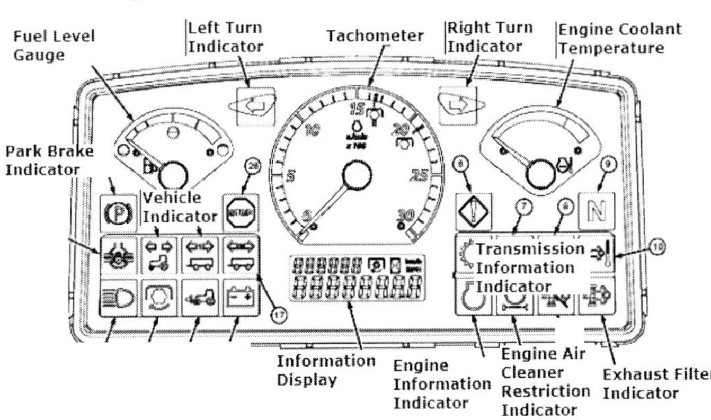

Nowadays, many tractors are equipped with latest dashboards where several light indicators are shown. Every indicator has LED light that glows on the dashboard against the working of the tractor. It is the duty of tractor operator to monitor and see these indicators while driving the tractor, as such indicators shows the entire setup of the tractor.

1.14. Check tightness of wheel nuts

It is the duty of the tractor operator to thoroughly examine the tractor wheels for any looseness or defectives parts as such faults can be dangerous.

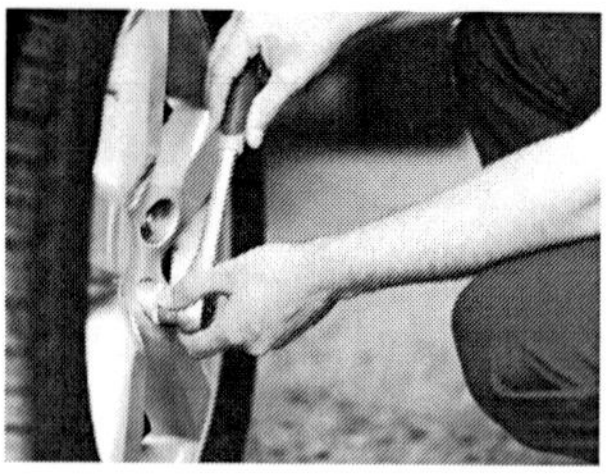

Loosening of tractor wheel bolts happens due to loosen over time as wheels remain always in rotation and are subject to heavy stresses on the fields/road. Tractor wheel nuts can be tighten by using a torque wrench as per the size of the nut. Wheel nut should always be checked after the repair of tyre puncher.

1.14.1 Checking and adjustment of tractor wheel nuts:

- ❖ Put the tractor in the first gear and apply the handbrake.
- ❖ Use required torque wrench to screw the wheel nut.
- ❖ Set the torque wrench to the stipulated value.
- ❖ Tighten the loose wheel bolts one after the other in criss-cross pattern.
- ❖ Turn the torque wrench clockwise.
- ❖ Loose the torque wrench so that the spring does not slacken and the torque value settings remain correct for the future.
- ❖ If the spring is slack, the torque wrench will give way before reaching the set value.

It is the duty of the tractor operator to check and monitor the tyres and their nuts before heading out in the field the next day.

1.15 Apply nipple greasing (once in 50 hours for dry land and once in 10 hours for wet land recommended)

1.15.1 Lubricating Grease:

- ❖ Good lubrication of tractor parts ensures reduced maintenance, optimised operating costs and extended lifetime for equipments.
- ❖ Tractor parts should be greased properly for smooth running and increases working efficiency as dust will make the parts harder.

1.15.2 Nipple greasing

Nipples should be greased after every 50 hours of tractor running. Greasing of nipple should be done using grease gun. For greasing the nipples, follow the steps:

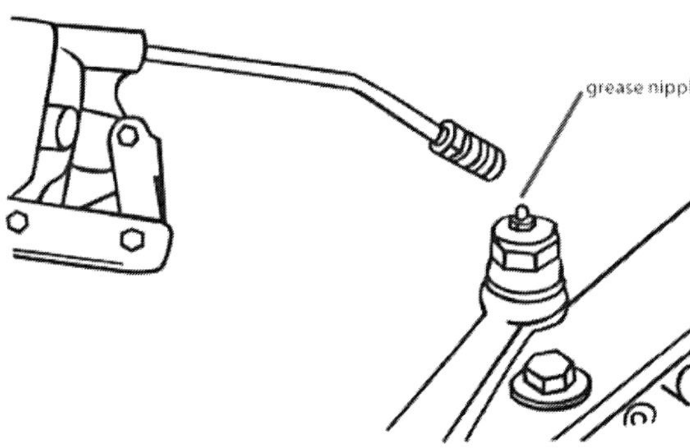

- ❖ Remove the dust and grease from the grease nipples before using the grease gun.
- ❖ Keep the grease gun directly in line with grease nipple and pump it slowly.
- ❖ Now remove the gun by moving it slightly on one side.
- ❖ Use the operator manual to lubricate the bearing, as over lubrication lead to rupture of seal which allows the dust to enter.
- ❖ Locate and change the grease nipple which is damaged or which does not allow the grease.

Chapter 2

Ballasting Importance and Process

2.1 Tractor Working

The basic guiding principle of tractor working is power = torque * angular velocity, so in case of tractors, there will be more angular velocity with high rate of rotation of wheels. Tractor with bigger tyres has more torque. It is noted that by applying similar power to a car, the car will happens to run faster than the tractor as it generates more torque.

Since the tractor has to work in field and need to pull the implements or trolley with loads, so they need the ability to pull, for which torque is required. It is noted that the entire engine construction and transmission is designed to have higher torque.

2.2 Ballasting

In tractor, torque can be increase by increasing the size of tyres. On adding the weight on the rims of the tractor wheel, the weight of the wheels can be increased. Such process of adding weight to the rear tyres or rear tyre rims is known as ballasting. Ballasting a tractor is an efficient method of increasing the weight of the tractor. There are many types of ballast which can be used to add weight to a tractor. Ballasting helps the tractor to reduce slipping of rear tyres and improves the pulling power.

Ballasting can be done in two ways:

- **Liquid ballasting:** It involves filling of the rear tyre with 80% water and remaining 20% with air.
- **Adding weights on tyre:** It involves adding weights on the tyre rim by bolting.

2.2.1 Selecting proper ballast

The following steps will help in selecting proper ballast while optimizing field speed and drive wheel slip which includes:

- Selecting relative high field speed, above 6 mph.
- Selecting an implement width which matches with this speed and available drawbar power of the tractor.
- Ballasting drive wheels to get a slip that is near the optimal slip for the particular soil conditions

Problems with over-ballasted tractor

- ❖ Increases fuel consumption and reduction in productive field output.
- ❖ Increases mechanical wear as a result of higher torque loadings in the tractor's drive train.
- ❖ Increases soil compaction.

Problems with under-ballasted tractor

- ❖ Potential for increased wheel slip that reduces field output.
- ❖ Increased tyre wear in certain conditions.
- ❖ The creation of deep tyre furrows in very wet conditions.

2.2.2 Types of weights

In tractor, weights can be added either at the front or at the rear which can be decided as per the implement and on the application of the tractor. Weight can be:

Cast iron weights

Cast iron weights are the most convenient which can be changed as per the ballasting needs.

Concrete weights

Concrete weights can be also used to replace cast iron,

Adding fluid

- ❖ Fluid such as calcium chloride solution can be added to the drive tires which is more economical than cast weights.
- ❖ The amount of liquid ballasting which is added, will depends on the size of the tires and the concentration of the calcium chloride solution.
- ❖ There are three different concentrations of solution that can be used and amount of ballasting in such case will depend on the concentration.

- ❖ While adding the fluid, it is noted that the tyre should not be overfilled with solution.
- ❖ Leave some air space for liquid volume changes.

2.2.3 Adding Weights

- ❖ Adding of weights should be done as per weight calculation.
- ❖ If the weight exceeds the required level, in such case, it may leads to more fuel consumption, soil compaction, damages in transmission.
- ❖ If the implement is at the rear end, it causes the front wheel to lift, so in such case, the weight should be added at the front.
- ❖ If front end loader is attached, it causes an imbalance at the rear portion of tractor, so in such case, the weight should be added at the rear.

Weight distribution for various tractor types

Tractor type	Front weight	Rear weight
2WD	25%	75%
FWA	40%	60%
4WD	60%	40%

2.3 Need of ballasting

It is observed that ballasting in tractor is mainly required at the time of pudding where soil is mixed with water in the fields. Apart from ballasting for pudding operations, there are many other operations which require ballasting in tractor like deep ploughing etc.

2.3.1 Ballasting on front axle

When a tractor pulls a heavy load, in such case, the transfer of weight is done from front axle to rear axle, due to which, tractor front wheels gets left up the ground and here is a possibility that a tractors gets overturn.

To avoid such situation, weight is added on front axle which increases the longitudinal stability of tractor while pulling heavy loads.

2.4 Ballast set-up process

Adding the required ballasting weight and keeping it correctly needs:

- ❖ consideration of vehicle and operator weight
- ❖ optimal distribution of weight between axles
- ❖ tyre pressures

- mounted implements
- travel speed

Processing Steps:

The process of ballasting involves collecting data which is particular for a tractor, mounted equipment and operations. There are certain steps which need to be followed by tractor operator for ballasting his tractor which covers:

Consulting the owner manual:

Tractor operator should be required to refer to owner's manual for correct process documentation as the tractor manufacturer correctly highlights the important instruction, guidelines for proper tractor set-up.

Analysing goal weight

Goal weight is the ball park gross weight which is ideal for your tractor and its operations. Before undertaking the process of adding or removing ballast, goal weight should be analysed. It is actually the gross weight of the tractor, ballast, operator and any extra load. Required gross weights for given loads and ground speeds are provided by manufacturers which is highlighted in tractor's manual.

In the table below shows the optimal gross tractor weights for given tractor types, with PTO power and operating speed.

Table: Optimal gross tractor weight per PTO power (kg/kW) at various operating speeds.

Tractor Type	<7 Km/Hr	8 Km/Hr	>9 Km/Hr
2WD & mfwd	80	75	65
4wd	65	60	55

Determine weight distribution

Weight distribution is important for performance and safety. Optimal distribution will vary for different tractor types, hitch points and mountings. Table below shows general recommendations which need to be followed.

Table: Front-to-rear axle ratio as a percentage of total weight

Tractor Type	Towed/Drawbar	Semi-	mounted		fully mounted	
	Front	Rear	Front	Rear	Front	Rear
2WD	25%	75%	30%	70%	35%	65%
MFWD	35%	65%	35%	65%	40%	60%
4wd	55%	45%	55%	45%	60%	40%

Measure current weight on each axle

Before setting the operating weight of a tractor, tractor operator should confirm the actual weight of the machine. The published weights of the tractor as provided by manufacturer are unlikely to exactly represent the tractor's actual specifications and options. In addition, mounted equipment contributes to the gross weight and its distribution. It is important to obtain separate measures for the weight on each axle and make sure that measurements are made while such as:

- fuel tank is at half capacity,
- operator is in the seat of the cab,
- main implements are mounted.

There are many opinions about how exactly tractor operator needs to be when factoring in mounted items while establishing goal weight. It is important to realise that measuring current weight on each axle can only be done once the requirement of ballast is done.

Add/remove appropriate weight

As per the recorded goal weights, tractor operator can add or remove the ballast from each axle. He/she should be aware of the weight limitations that are inherent to the type of tyres on the tractor. The ballast can be added by using cast iron weights or by filling the tyres with fluid.

Fine-tune traction by tyre pressure

Tractor operator has to set a correct goal weight for the tractor by using the tyre pressure which will fine-tune the set up. Adjusting the tyre pressure is easier and most effective way to get an optimal slip to grip ratio for a given surface and load conditions.

Chapter 3

PC & DC levers or Hydraulic Lift Performance

3.1 Hydraulic System

There are three important methods in hydraulic control system:

- Position control
- Draft control
- Mixed control

Position control:

In this, constant depth of ploughing is maintained by automatic adjustment of tractor draft. Here, the control valve gets operated directly by the driver to raise the lower or hold an implement, mounted on the linkage at any selected height.

Draft control:

In this, working depth of any implement gets controlled without the need for a depth wheel on the implement. The hydraulic control valve reacts to changes in the loading in either the top or lower links whose arc due to changes in the draft or pull is required by the implement. If any implement goes too deep, in such case, its draft increases which is sensed through the top link or lower links. The control system in such case will raise the implement till the draft is back to the present level and the implement is at the original depth again using the draft control system.

3.2 PC and DC Levers

The Draft Control system uses the top link signal for raising or lowering the implement working depth to maintain a constant draft. The system has two control levers:

Levers

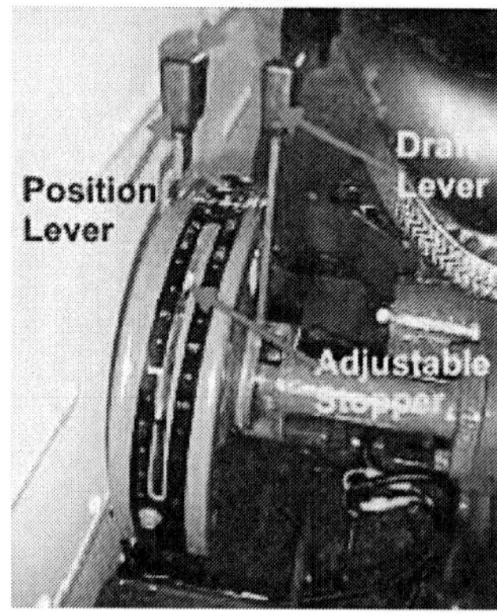

- ❖ Draft Control Lever
- ❖ Position Control Lever

The basic functions of the levers are:

- ❖ Draft Control
- ❖ Position Control
- ❖ Draft Control with Position Control (Blending)

3.2.1 Position Control Lever

- ❖ The main lever position is directly related to the height of the three-point hitch.
- ❖ It is most widely used and should be used for all implements carried in three-point hitch above ground.
- ❖ In this mode, three-point hitch maintains constant height depending only on the position of the main lever.

3.2.2 Draft control Lever

- ❖ The main lever actually selects the tractive force of tillage implements.
- ❖ It should only be used for tillage.
- ❖ In this, unit adjusts the height of the hitch so the traction force remains constant.
- ❖ The amount of tractive force is adjusted by moving the main lever forward increases the tractive force, or plowing depth.

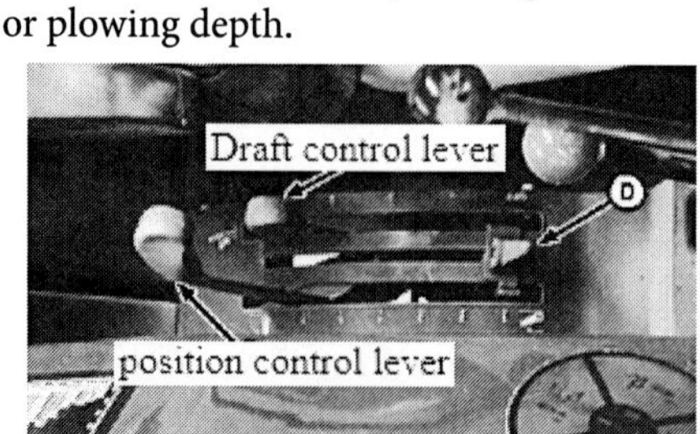

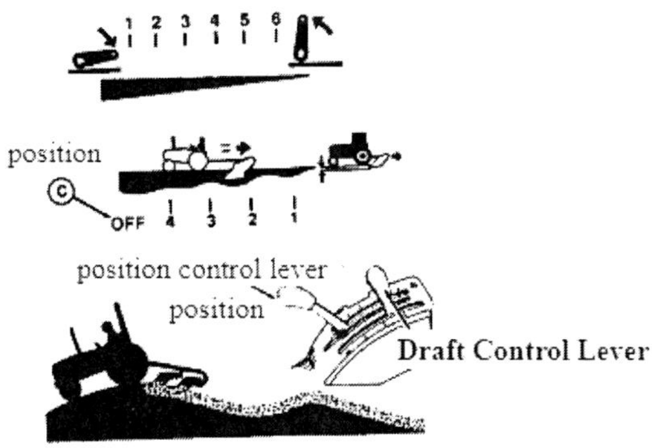

Draft control lever controls the amount of load required before the hitch responds.

With lever placed fully forward to the position marked "off", there is no draft sensing.

Placing the lever toward the rear position will reduce the amount of draft load required to override the position setting set by the position control lever and raise the rockshaft.

Using draft load sensing:

- ❖ The implement will raise and lower to follow the ground contours while maintaining a nearly constant depth.
- ❖ The implement is raised slightly to get through tough spots so you do not have to shift to a lower gear.

3.2.3 Draft Control with Position Control (Blending)

In certain conditions, such as cultivating at shallow depths, it is desirable to limit the working depth of the implement.

3.3 Operate PC & DC levers as per the need and in combination for mixed control/blending

The Inner Quadrant has three ranges:

- ❖ Position Control
- ❖ Constant Pumping
- ❖ Pressure Control

An Inner Quadrant Lever performs a different function in each of the three ranges.

3.3.1 When to use Position Control

- ❖ Position Control range is used for operating 3-point hitch mounted tools that are not Draft Controlled, like soil scoops, posthole diggers and cranes.
- ❖ Implements of this type may be held at a fixed height, depth or intermediate position, regardless of Draft loads, thus establishing an infinite number of lower link positions which correspond to the inner quadrant lever.

Using Position Control

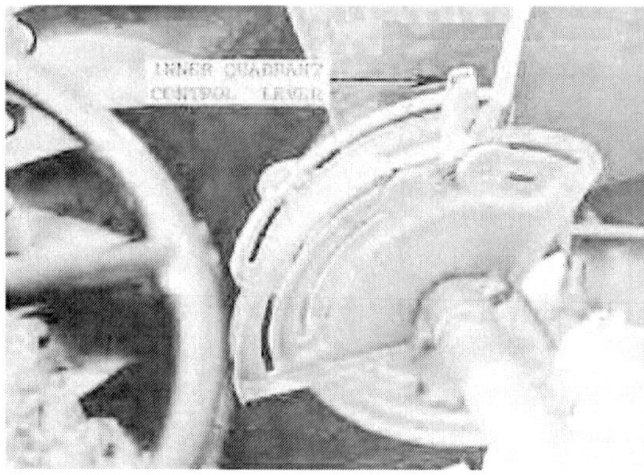

- ❖ When operating in the Position Control Range, place the Draft Control Lever in the rully raised position.
- ❖ Shift the inner quadrant lever to the Position Control Range.
- ❖ Establish a transport stop by moving the stop for the Pressure Control Range to the upper end of the slot which provides correct setting of inner quadrant lever for transport when operating in Position Control.
- ❖ The inner quadrant lever may be used to raise, lower and operate 3-point hitch implements that are not draft controlled.
- ❖ Moving the inner quadrant lever toward transport position will raise the implement.
- ❖ Moving the inner quadrant lever toward down position will lower the implement.
- ❖ Locating the inner quadrant lever at intermediate positions in the Position Control range which locate the lower links at corresponding intermediate positions.
- ❖ The quadrant has an adjustable stop that allows the operator to return the inner quadrant lever from transport to the same desired operating position.

3.3.2 When to use Draft Control Lever

- ❖ Draft Control is used when the 3-point hitch implement is of the soil engaging type and when the implement is required to maintain a uniform working depth and follow the contour of the ground.
- ❖ When using a Draft Controlled implement like plow, sub-soiler, etc. the draft system automatically maintains a furrow at implement working depth which transfers the weight of the tractor rear wheels to provide traction, regardless of field undulations.
- ❖ When working in a field that possesses various soil textures, uniform depth can be maintained by minor adjustments to the Draft control lever.

Using the Draft Control Lever

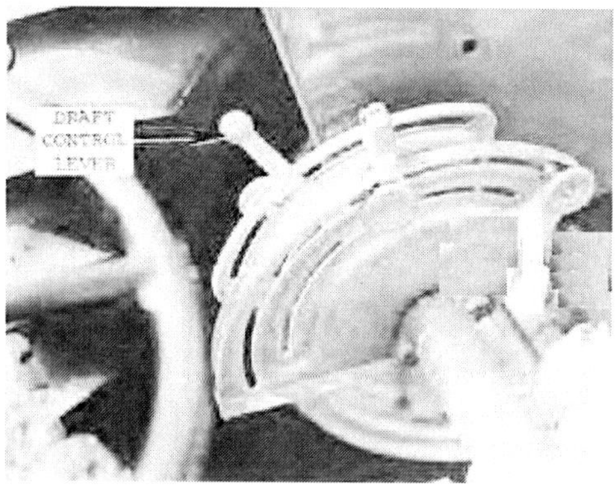

- ❖ The Draft Control Lever is used to raise, lower and regulate the operating depth of draft tools.
- ❖ When using the Draft Control Lever, place the inner quadrant lever in the transport position of the Position Control range.
- ❖ To lower a Draft Controlled implement from the transport position to the ground, move the Draft Control Lever down.
- ❖ If the implement is running too shallow, continue to move the Draft Control Lever down until the desired depth is obtained.
- ❖ If the implement is running too deep, move the Draft Control Lever up until the desired depth has been obtained.
- ❖ The adjustable locator may be set in line with the Draft Control lever to be used as a reference which allows operator to lower the implement to the same selected working depth.
- ❖ To raise a Draft Controlled implement to the transport position, move the Draft Control Lever up until the transport position is reached.

3.4 Hydraulic lift performance with weight fitted to lower links

- ❖ Lift capacity is the most deceptive gauge by which a tractor operator can compare tractors that are used in agriculture.
- ❖ Whether it is lift capacity of the loader or three-point lift, often times the number quoted is not the weight the tractor can actually lift.
- ❖ Lift capacity is measured by several different means as lift capacity to full height, static lift capacity, pivot pin capacity, bucket center capacity, lift arm capacity and 24 inches aft of lift arm capacity.

3.4.1 Hydraulic Lift Capacity Test

Hydraulic lift capacity and flow tests are conducted to determine the maximum lift capacity of a hydraulic system through the full lift range. While checking of hydraulic lift capacity, it should be kept in mind that unballasted tractor should support and secure in horizontal position at the rear wheel axis so that the tvres are not deflected by the reactive forces of the power lift. The linkage to be adjusted as appropriate for the tests with or without the coupled frame, to achieve typical and repeatable arrangements as follows:

- ❖ The linkage shall be adjusted in such a way as to achieve the power range required and a height of the lower hitch points in lower position from 200 mm to 230 mm for different categories.
- ❖ The height to be obtained with the loaded radius of tyre with inflation pressure as recommended for field operation.
- ❖ For tractors which do not achieve the standard power range, lifting force shall be measured at the maximum achievable power range.
- ❖ The upper link is to be adjusted to required length to bring the mist of the frame vertical when the lower links are horizontal.
- ❖ In case of more than one upper or lower link point, the points used shall be those specified by the manufacturer/applicant which includes in test report.
- ❖ In case more than one point attachment to connect lift rod to lower links, connection points to be used shall be as specified by manufacturer/applicant which should include in test report.
- ❖ The initial adjustments will able to cause the mast to rotate through a minimum of 10° over the full range of lift, or if not, then fact shall be stated in the test report.

3.4.2 Static Lift at Lower Hitch Points

In case of static lift at low hitch point, external vertical downward force to be applied to a horizontal bar that connects the hitch points. Such force can be measured which lies in the central longitudinal plane of the tractor that maintain vertical throughout the lift range. In such case, care should be taken to avoid torsional components in this force as it can affect the accuracy of measurements.

The lifting force and corresponding pressure of hydraulic fluid needs to be measured at 8 minimum of 6 points approximately equally spaced throughout the range of movement of the lift. At each point, the force shall be the maximum which can be exerted against a static load and the range of movement should be measured.

The minimum lifting forces should be corrected to pressure that corresponds to maximum hydraulic power. The corrected value shall constitute the maximum vertical force which can be supported by the hydraulic lift throughout its full range of movement.

Chapter 4

Gear & RPM selection

4.1. Gears

- Gear is a toothed wheel designed to transmit torque to another gear or toothed component.
- Teeth of a gear are shaped to minimize wear, vibration and noise, and to maximize the efficiency of power transmission.
- Different sized gears are used in pairs for a mechanical advantage, which allows the torque of the driving gear to produce a larger torque in the driven gear at lower speed, or smaller torque at higher speed.
- The larger gear is known as a wheel and the smaller as a pinion.

4.1.1 Types of gear:

Spur gears:

- These gears are flat with teeth projecting radially out in the plane of the wheel, straight-cut gears.
- These gears can be fitted only to parallel axles.

Helical gears:

- ❖ In this, teeth are cut at an angle, which allow for more gradual and smoother meshing between gear wheels by removing whine characteristic of straight-cut gears
- ❖ In double helical gears, the teeth are V shaped.
- ❖ Each gear in a double helical gear can be thought of as two standard, but mirror image, helical gears stacked.

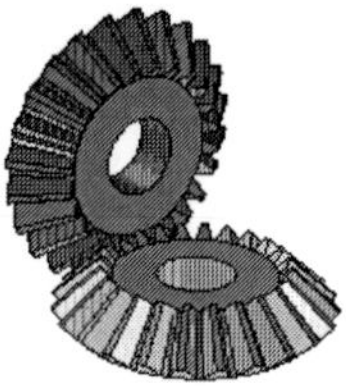

Bevel gears:

- ❖ These gears allow to change the operating angle.
- ❖ In this, a single wheel is designed to work with it's complementary wheel and no other.
- ❖ The combination of four bevel gears makes a differential gear that transmit power to two axles spinning at different speeds like on automobile.

4.1.2 Gears used in tractors

Tractors need more torque to work in the field, this traction is given by reducing engine speed and needs varied PTO (Power Take Off) speed for various applications. The following are some of the common transmission system used in tractors.

Manual Transmission

In manual transmission both the clutch and gears are operated manually by the driver. Normally there are 6-8 forward and 1-2 reverse gears.

1. Sliding gear
2. Constant mesh
3. Synchronized or Synchromesh
4. Power shift

Automatic Transmission

1. Hydrostatic
2. Hydro-dynamic

Sliding gear transmission

Sliding gear transmissions has simple arrangements of spur gears and shafts. It has input shaft and output shaft which are held parallel to each other in a housing or case while the third shaft called as idler shaft reverses the power to output shaft.

In each gear pair, one gear is located on input shaft while other on output shaft. Gears on input shaft is attached to shaft by splines or keys. Gears on output shaft slide which engage mating gear on input shaft. Reverse occurs when sliding gear engages idler gear.

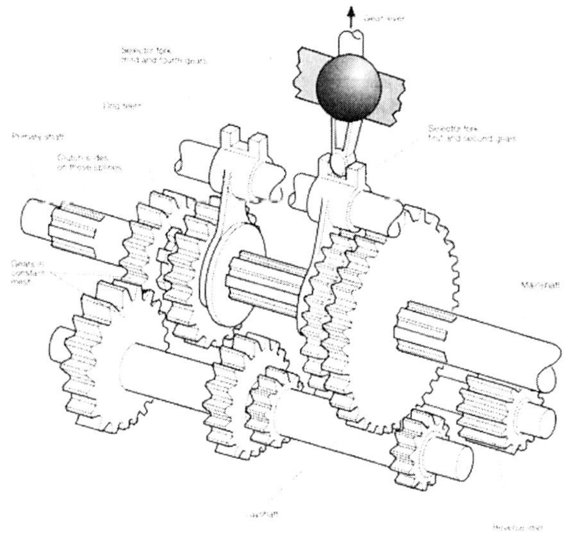

Constant mesh type

Such gears are always in mesh and uses helical gears. Gears are engaged by shifting couplings, which slide along the splines on the counter-shaft and the output shaft of the gear box.

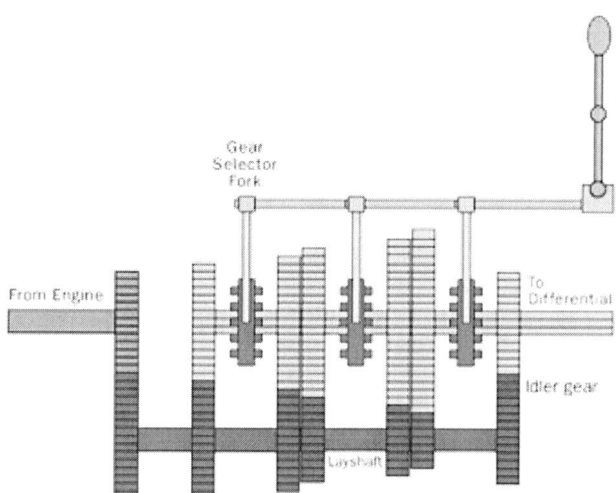

Simple unsynchronized systems:

In this, gears are spinning freely and their relative speeds gets synchronized by the operator to avoid noisy which damages clashing and grinding when trying to mesh the rotating teeth.

Synchronized systems:

If the gearbox is of constant mesh type where all gears are in mesh but only one of these meshed pairs of gears is locked to the shaft on which it is mounted at any time, the others being allowed to rotate freely, hence reducing the skill required to shift gears. Such systems that will automatically mesh while changing gears will be same as manual shift except it has synchronizers that can be shifted on the go. The clutch should always be used for starts, stops as well as shifting. It usually has 6-8 forward gears and 1-2 reverses.

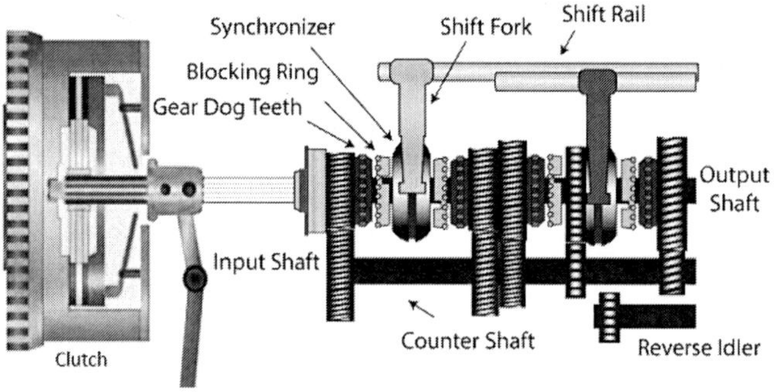

Hydrostatic Transmission:

Hydrostatic transmissions transmit all power with hydraulics using power of oil. The greatest advantage of a hydrostatic transmission is ability to infinitely vary the ground speed and quickly change the directions. Hydrostatic is, the best choice for turf mowing applications or for any tasks that require constant speed and direction changes within a small area.

Continuously variable Transmission:

It is a different type of automatic transmission which smoothly changes its gear ratio by varying the diameter of a pair of belt or chain-linked pulleys, wheels or cones. Its designs are usually as fuel efficient as manual transmissions in city driving, but early designs lose efficiency as engine speed increases.

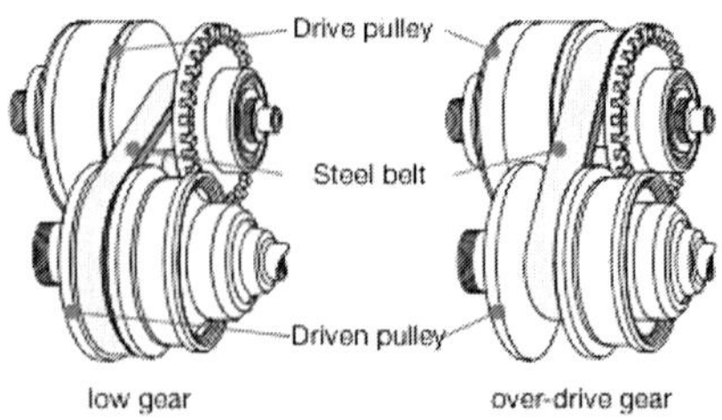

4.1.2 RPM

- RPM is Revolutions per Minute.
- In Engine, RPM tells about how many times the shafts have completed its rotation with respect to fixed axis in a minute.
- It shows circular motion, which happens as an object as turns around an axis.
- RPM is useful speed measurement as it tells how quickly you can rotate when not under a load.
- If RPM Of engine increases, torque increases to certain value of RPM while if RPM goes on increasing, in such case, torque starts decreasing.

4.2 Proper selection of Gear & RPM

4.2.1 Proper Gear Selection

Tractor is a traction machine which is required to pull a load. It works more efficiently in a higher gear with a reduced throttle setting. When pulling a light load, it uses higher gear and reduces throttle to maintain the desired ground speed which is difficult when using the PTO (power take-off) as many PTO driven machines is operated at rated speed to work properly. Engine speed/PTO speed is often shown on tractor instrument panel.

For better performance and durability of a tractor during field operations, it is necessary to select the optimal gear setting for the operation. Many older tractors on farms today have manual shift gear drive with four speed having high, medium and low range which gives operator 8 or 12 gear speeds to use.

The power shift tractors gets shifted on fly with just a push of a button. Such tractors can have as many as 24 forward speeds and the same number for reverse. In tractor, more gears allows for more selection in speed in small increments. If the tractor is moving down on the hill, then use the tractor gear as in such situation, the tractor engine serves as a brake. If the tractor operator has a doubt about the type of gear to be used then in such case, he/she should select the lowest-speed gear and shift before starting down the hill.

4.2.2 Proper RPM Selection

Tractor is a self-propelled power unit having wheels or tracks for operating agricultural implements and machines including trailers. A tractor is used mostly for plowing and disking which carries many variables that will judge soil type, pulling of implements and time for specific work. In order to match the tractor with the type of job or type of tool to be used for particular work, the PTO horsepower, engine type and rpm are important.

Rpm is used to give the engine operating speed. The rpm must be considered when engines are applied to equipment driven by a rotating shaft. The piston speed, which is indicator of engine internal forces and estimator or engine life is useful in determining the limit of rpm for efficient and economical operation when the limit is not specified by the manufacturer.

Types of cutter and slasher:

- ❖ For using five-foot cutter, a 20 PTO horsepower engine is used.
- ❖ For using six-foot cutter, a 30 PTO horsepower engine is used.
- ❖ For using 90–100cm slasher, 18 – 22 horsepower engine is good.
- ❖ For using 120cm slasher, 25 – 28 horsepower engine is good.
- ❖ For using 130cm slaster, 28 horsepower engine is good.
- ❖ For using 150cm slaster, 30 – 40 horsepower engine is good.

Types of horsepower:

- ❖ Tractor with 35 horsepower of engine speed has 2300 rpm.
- ❖ Tractor with 40 horsepower has four stroke water cooled engine where bore and stroke of engine is 95 mm and 110 mm that has 2100 rpm.

Types of engine:

- ❖ Gasoline engine rpm; 1500 – 2500
- ❖ Diesel engine rpm; 1500 - 2500
- ❖ Propane engine rpm; 1500 - 2400

4.3. Check the functioning of different gauges in the instrument panel such as RPM gauge, hour meter, fuel gauge, battery charging indicator, air filter choke indicator, etc

4.3.1 RPM gauge

RPM gauge shows the speed of engine spinning. It is revolutions per minute which shows how fast the engine turns. Knowing the engine RPMs will help in:

- ❖ Deciding, when to shift gears on manual transmission
- ❖ Increasing the mileage by shifting gears at correct RPM level
- ❖ Determining, if engine and transmission are operating properly
- ❖ Driving vehicle without damaging the engine

Measuring RPM

- ❖ To measure the RPMs of your tractor, you have to use a handheld tachometer which will measure the vehicle's RPMs.

- Tachometer usually shows single or double digit numbers that increase into a red band, which shows RPMs above the redline for the vehicle.
- The redline shoed the maximum number of RPMs which your vehicle can easily manage.
- When you start the tractor, you will find that the tachometer climbs rapidly, before settling on the number of RPMs as prescribed.
- Now press the speed pedal and see its effect on tachometer. You find that on increasing the amount of air in the engine, the RPMs climb to match.
- For frequent change in tachometer reading, run your tractor at different speed.

4.3.2 Hour meter

Hour meter is a gauge/instrument which tracks and records the elapsed time shown in hours and tenths of hours. It is used to log the running time of equipment to assure proper maintenance of expensive machines or systems which can be in terms of replacing, changing or checking parts, belts, filters, oil, lubrication or running condition in engines, motors, blowers, fans etc.

The reason for using an hour meter is for maintenance. If you rely on the actual running time of an engine to schedule maintenance from the last service, rather than calendar days or miles driven, there is less guesswork and greater correlation to the true operating life of the equipment.

On the dashboard where all meters are installed, see for small rectangular 8-digit odometer. Newer tractors have digital hour meter that is displayed on the dashboard. Read the hours on the hours odometer having 6 digits, a decimal point and then 2 digits for a partial hour reading.

4.3.3 Fuel gauge

In tractor, fuel gauge is an instrument which is used to indicate the amount of fuel in a fuel tank.

It comprises of two parts:

- Sending unit, which is in the tank
- Indicator, which is located on dashboard

Sending unit

- Sending unit uses a float connected to potentiometer with ink design.
- As the tank empties, the float will drop and slides a moving contact with a resistor and increases resistance.
- When the resistance is at a certain point, it will turn on the low fuel light.

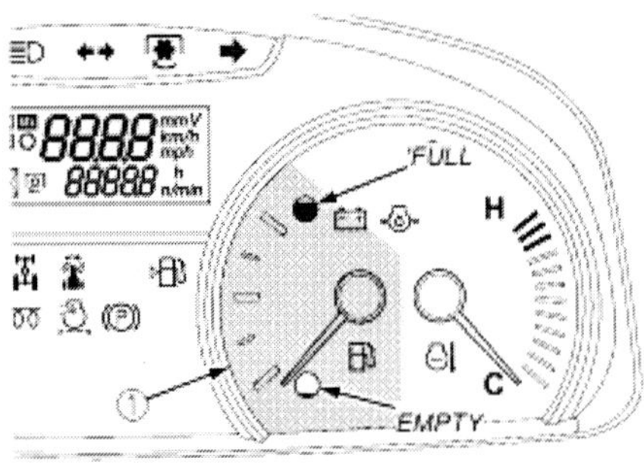

Indicator unit

- Indicator unit is mounted on the dashboard which measures and displays the amount of electric current flowing through the sending unit.
- When the tank level is filled and maximum current flows, the needle points to 'F' showing full tank.
- When the tank is empty and less current flows, the needle points to 'E' which shows empty tank.

4.3.4 Battery charging indicator

The tractor battery has its own built in recharging system which include the alternator and voltage regulator to keep the battery charged and provide electricity to the engine while it is running.

The charge indicator is an ammeter that shows whether the alternator or generator is charging the battery properly or not. When the operator start the tractor every time, the battery is discharged, but during its operation, the battery gets recharged again. Gauges will show + or - charge. The battery indicator lights will show red when it is less charged. If the battery is discharging, the engine will not start next time.

The battery indicator light on tractor dashboard gives indication to tractor operator that something is wrong with the tractor charging system which could be due to broken alternator belt or failing of alternator altogether.

4.3.5 Air filter choke indicator

Internal combustion engines needs a good amount of air to keep themselves running. At the same time, things like dust and pollen, in the air are not good for the tractor engine. For this, the air filter is required which collects any debris floating in the air which saves these debris from getting inside the engine. With the matter of time, all collected debris will clog up the filter and reduces airflow in the engine which further reduces engine performance.

In tractor dashboard, the chocked air filter indicator light warns the tractor operator of reduced airflow in the engine. Tractor operator should note that when such type of indication appears on the dashboard, in such case, he/she has to replace the air filter or can get clean up on inspection. After the air filter is replaced, he tractor operator has to manually turn off the warning light using a reset button which is present on the dashboard of the tractor.

Chapter 5

Implements and Trailer

5.1 Tillage implements

Tillage is a mechanical manipulation of soil to give favourable condition for crop production. Soil tillage has breaking compact surface of earth to certain depth and to loosen the soil mass, to allow roots of the crops to penetrate and spread into the soil.

Tillage can be:

- Minimum Tillage: It is minimum soil manipulation needed to meet the tillage requirements for crop production.
- Strip Tillage: It is a tillage system where only isolated bands of soil are tilled.
- Rotary Tillage: It is the tillage operations using rotary action which will cut, break and mix the soil.
- Mulch Tillage: It is used for preparations of soil where a plant residues or other mulching materials are left on or near the surface.
- Combined Tillage: It operates by using two or more different types of tillage tools or implements so as to simplify, control and reduce, number of operations over a field.

5.2 Classification of tillage implements

Tillage implements are classified as:

5.2.1 Primary tillage implements

These are the implements that are used for opening and loosening of the soil such as ploughs.

Wooden plough

- It is an implement which is made of wood with an iron share point.
- It consists of body, shaft pole, share and handle.
- It is drawn with bullocks.
- It cuts a V shaped furrow and opens the soil but there is no inversion.
- With this, ploughing operation is not perfect as some unploughed strip is always left between furrows.

Soil Turning Ploughs

- ❖ It is made of iron which is drawn by a pair of bullocks or two depending on the type of soil.
- ❖ It is drawn by tractor.

Mouldboard Plough

- ❖ It comprises of frog or body, mouldboard or wing, share, landside, connecting, rod, bracket and handle.

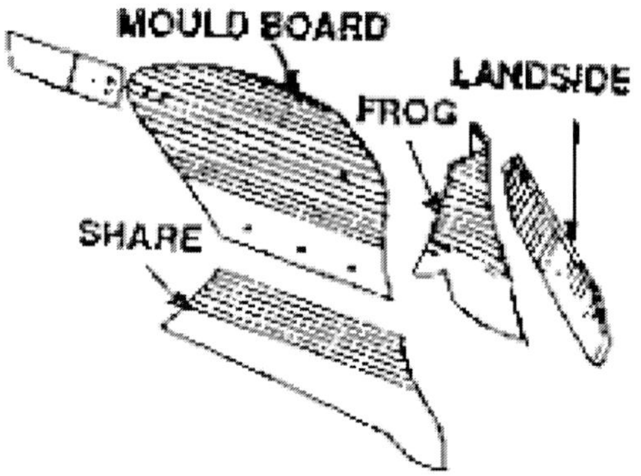

- ❖ Such plough leaves no unploughed land as the furrow slices are cut clean and inverted to one side resulting in better pulverisation.
- ❖ The animal drawn mouldboard plough is small with plough depth of 15 cm, while two mouldboard ploughs that are bigger in size are attached to the tractor and ploughed to a depth of 25 to 30 cm.
- ❖ These ploughs are used where soil inversion is required.
- ❖ Victory plough is an animal drawn mouldboard plough having shorter shaft.

Disc Plough

- ❖ It is slightly similar to common mouldboard plough.
- ❖ It has a large, revolving, concave steel disc that replaces the share and the mouldboard.
- ❖ The disc turns the furrow slice to one side with a scooping action.
- ❖ The size of the disc is around 60 cm in diameter which turns 35 to 30 cm furrow slice.
- ❖ It is good for land in which there is much fibrous growth of weeds since disc cuts and incorporates the weeds.
- ❖ It works good in soil which is free from stones.
- ❖ It needs no harrowing to break the clods of upturned soil as in case of mouldboard plough.

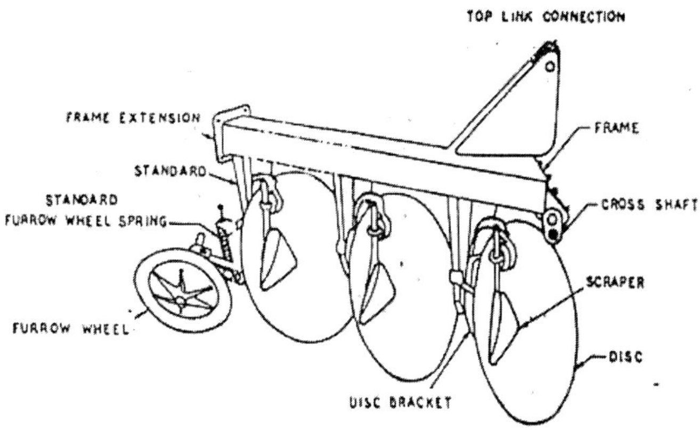

One-way Plough

- The plough bottom in this is hinged to the beam such that the mouldboard and the share can be reversed to the left or right side of the beam.
- In this, the adjustment saves the trouble of turning the plough in hilly tracts, but shows inversion of furrow slice to only single side.

Special Ploughs

Subsoil Plough:

- It is designed to break up hard layers or pans without bringing them to the surface.
- Its body is wedge shaped and narrow while the share is wide so as to shatter the hard pan and making only a slot on the top layers.

Chisel Plough:

- It is used for breaking hard pans and for deep ploughing with less disturbance to the top layers.
- Its body is thin with replaceable cutting edge to have minimum disturbance to top layers.
- It has replaceable share to shatter the lower layers.

Rotary Plough/Hoes:

- It cuts the soil and pulverizes it.
- In this, cutting of soil is done by either blades or tynes.
- The blade types are mostly used.
- The depth of cut is up to 12 to 15 cm.
- It is suitable for light soils.

Basin Lister:

- It is a heavy implement with one or two mouldboards or shovels.
- Such shovels are mounted on special type of frame on which they act alternately.
- It is used to form listed furrows to save free runoff of rainfall and blowing off the soil in low rainfall areas.

5.2.2 Secondary Tillage implements

Secondary tillage implements are used for:

- removing the weeds by cutting.
- loosening the soil and facilitate the moisture entry into the soil.
- smoothening and levelling field process.
- breaking the clods and then pulverize for efficient seed bed preparation.
- cutting the crop residue like straw and further mix them with soil for faster decomposition.

There are different types of secondary tillage implements such as:

- Cultivators
- Harrows
- Planks and rollers

1. Cultivator:

- It is an implement which is used for finer operations such as breaking clods and working the soil to a fine tilth while preparing seedbed.
- It is also known as tiller or tooth harrow.
- It is used to further loosen the previously ploughed land before sowing.

- It destroy weeds that germinate after ploughing.
- It has two rows of tynes which are attached to its frame in staggered form which provides clearance between tynes so that clods and plant residues can freely pass through without blocking.
- In this, the provision is made in the frame by drilling holes so that tynes can be set close or apart as desired.
- The number of tynes in this, ranges from 7 to 13.

- The shares of the tynes can be replaced when they are worn out.

Sweep Cultivator

- It is an implement which is used to prepare the land using ordinary implements.
- It comprises of large inverted V shaped blades attached to a cultivator frame.
- In this, the blades run parallel to soil surface at a depth of 10 to 15 cm which are arranged in two rows and staggered.
- It is used to cut up to 12 to 15cm depth of soil during first operation after harvesting and shallower during subsequent operations.
- It works frequently to control weeds and for harvesting groundnut.

Field Cultivator:

- It has a metal teeth to break the soil.
- It is a tractor powered agricultural equipment which is used to complete tillage operations in many types of arable crop fields.
- It mingles the soil while overthrowing weeds, destroying them but leaving them in the soil to limit soil erosion.
- Its main work is to prepare proper seedbed for the crop to be planted in.
- It ensures that the growing crop has enough water and nutrients to grow well during the growing season.
- In this, hydraulic cylinder raise and lower the implement and control its **depth.**

Row Crop Cultivator:

- It is used for weed control between the rows of an established crop.
- It is raised and lowered by three-point hitch and its depth is controlled by gauge wheels.

Hand-action Cultivator:

- It is a type of push-pull cultivator having rotating tines attached to a handle.
- It helps in maintaining gardens which have already been cultivated **and weeded in rows.**

Hand Cultivator:

- It is made of metal or synthetic material.
- It is used with a chop and pull motion that allows three curved prongs to catch and remove weeds or to break up dirt clods.

2. Harrows

Harrows are used for shallow cultivation in operations such as preparation of seedbed, covering seeds and destroying weed seedlings. Harrows are of two types:

- disc harrow
- blade harrow

Disc Harrow

- It comprises of number of concave discs of 45 to 55 cm in diameter.
- The discs are smaller in size than disc plough with more discs arranged on a frame.
- The discs are fitted 15cm apart on axles.
- In this, two sets of discs are mounted on two axles.
- Here all discs revolve together with axles.
- The discs cut through the soil and effectively pulverise the clods.

Blade Harrow

- It is used for different works like removing of weeds and stubbles, crushing of clods working of soil to shallow depth, covering the seeds, intercutivation and harvesting of groundnut etc.
- It is useful for intercultivation.
- It is of two types; indigenous and improved.

3. Plank and Roller

Plank

- It is a simple implement comprises of heavy wooden beam of length 2 m.
- In this, the shafts and handle are fixed to the beams.
- On using it, many clods are crushed due to its weight.
- It helps in micro levelling and slight compaction required after sowing.

Roller

- It is used for crushing the hard clods and to compact the soil in seed rows.

- It is used for flattening land or breaking up large clumps of soil, after ploughing or disc harrowing.
- A heavy roller has one or more cylinders made of thick steel, a thinner steel cylinder filled with concrete, or a cylinder filled with water.
- In a water-filled roller, water gets drained out for lighter use or for transport.
- They prepare optimal seedbeds by making them as flat as is practical and moderately firmed.
- They get trailed after other equipment like ploughs, disc harrows or mowers.

Implements for Layout of seedbed

- Country plough and ridge plough are used for laying out the field in ridges and furrows or to layout irrigation channels.

- Ridge ploughs, when attached to a frame can be used *for* making broad-bed *furrows.*
- Bunds for irrigation in garden lands are made by manual labour using spades.
- Bunds are formed across the contours in low rainfall regions to conserve soil moisture.
- The bund consists of pair of iron mould boards fixed in opposite direction facing each other with the front end opening outwards and rear and closing in to form bunds.
- Marker is used to mark intercepts for transplanting seedlings by square planting method.
- It consists of a beam to which 3 or 4 wooden tyres are fixed, the spacing of which depends on the spacing the crop.
- When running in two directions, very shallow furrow markings are formed.
- Seedlings are transplanted at the intercepts.

Implements for sowing

Plough

- In this, seeds are dropped by hand in the furrow formed by country plough.

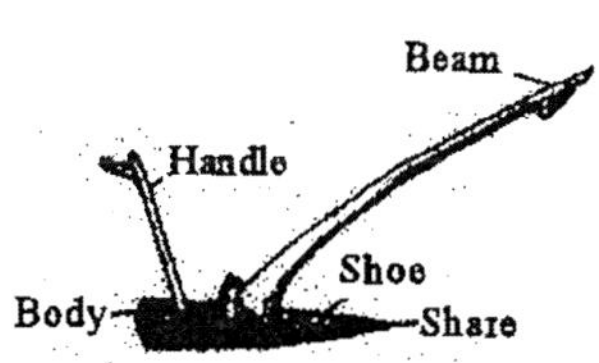

- The seeds fall at uneven depths due to falling at random in furrow slice.
- To avoid falling of seeds at uneven deph, *Akkadi* is used which is a hollow bamboo tube that is sharpened at one end with wide hopper at other end which is tied to country plough with rope that drops seed in *akkadi' s* hopper.
- In akkadi hopper, seed passes through the tube and fall in the furrow opened by the plough.

Seed Drill

- It consists of wooden beam to which 3 to 6 tynes are fixed.
- The tynes in this, open the furrows where seeds are dropped.
- In this, holes are made in tynes where the bottom ends of bamboo or metal seed tubes are fitted.
- The seed tubes are connected at the top to a wooden seed receptacle called as hopper.
- The seeds are fed at uniform rate in hopper using skilled labour that walks behind the seed drill.

Implements for intercultivation

- **Country plough and ridge ploughs** are used for earthing sugarcane, potato etc.
- Country plough runs to shallow depth in order to control weeds in widely spaced crops and fruit trees.
- **Small sized blade harrows** are widely used for intercultivation.
- Inter-cultivation in close growing crops is done with *danthi* which covers one row only.

- Depending on inter-row spacing of crop, blade length is 10cm which is less than inter-row spacing of crop.

- **Tobacco blade harrow** has longer blade than its beam so as to scrape the weeds on the soil without damaging the brittle petioles of tobacco.
- **Star weeder** is small implement that is used by the labour having long wooden or iron vertical rod with small horizontal rod for holding the implement.
- On the other end of s**tar weeder**, two star like wheels and a small blade of 10 cm are attached.
- In this, pointed teeth of rotating wheels loose the soil and helps in movement of an implement while the blade helps in cutting the weeds.

5.2.3 Leveling equipment

Leveling Blade

- It is an agricultural implement that levels the field much better than a labourer
- It is a simple and rugged equipment that can be directly fitted to the tractor

- It also levels the land of construction sites, playgrounds, horse riding clubs, etc
- It is a heavy equipment which is designed particularly for leveling purposes
- It is easy to maintain and use
- It levels the field accurately
- It saves time and labour needs
- It gives no extra load on the tractor
- It can be used in wet or dry condition of the soil
- It adjusts the depth of tilling as per the needs

Laser guided land leveler

It is used for leveling the field in certain degree of required slope by using guided laser beam in the field.

It is a cheapest way of leveling the land which is not cumbersome and involves less time.

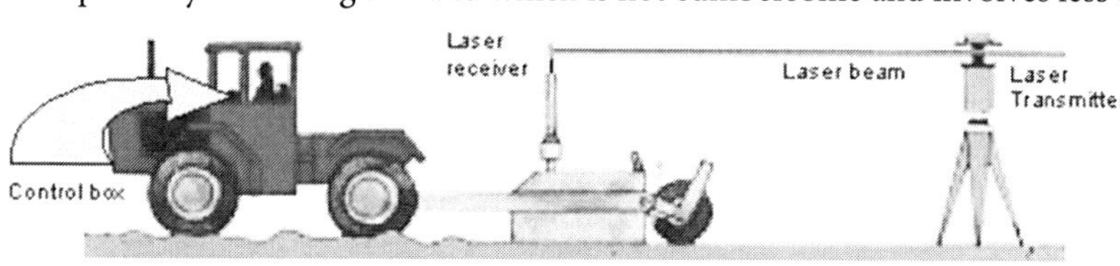

Parts:

- ❖ Drag scraper/bucket is a bucket which is either a 3-point linkage which is kept on or is pulled by a tractor. It is easier to connect the tractor's hydraulic system to an external hydraulic by 3-point-linkage system.
- ❖ Laser transmitter is kept on a tripod that allows the laser beam to sweep above the field.
- ❖ Laser receiver is a multi-directional receiver which detect the position of laser reference plane and transmits signal to the control box.
- ❖ Control box accepts and process the signals from the machine which is kept on the receiver that shows drag buckets position relative to the finished grade.
- ❖ Hydraulic system of tractor supply the oil to raise and lower the leveling bucket.

Working:

- ❖ Laser guided land leveler system includes a laser-transmitting unit that emits an infrared beam of light which can travel up to 700m in a straight line.
- ❖ In this, the receiver senses the infrared beam of light and converts it to an electrical signal.
- ❖ Here, the electrical signal is directed by a control box to activate an electric hydraulic valve.
- ❖ The hydraulic valve raises and lowers the blade of a grader to keep it following the infrared beam.
- ❖ Laser leveling of a field is done using dual slope laser which automatically controls the blade of the land leveler for grading the surface in order to remove all undulations that holds the water.
- ❖ In this, laser transmitters create a reference plane over the work area by rotating the laser beam at 3600.

❖ The receiving system detects the beam and guides the machine to maintain proper grade.

❖ The laser will level or sloped in two directions which can be done automatically without the operator touching the hydraulic controls.

5.3 Trailer

5.3.1 Introduction

❖ Trailer is a transport vehicle used for carrying agricultural and forestry products or articles.

❖ It is intended for being coupled to an agricultural tractor.

❖ It transfers a part of its total load on the towing vehicle and carries the rest of the load in its axle.

5.3.2 Types

❖ Agricultural Balanced Trailer: It is an agricultural trailer whose total load is supported by its wheels when detached from the towing vehicle. It is a double or multiple axle trailers.

❖ Fixed Platform Trailer: It is an agricultural trailer in which the platform is fixed with the chassis.

❖ Tipping-Platform Trailer: It is an agricultural trailer in which the platform can be tilted.

5.3.3 Features

❖ The height of the trailer with tyres inflated at pressure as recommended by manufacturer when measured from the level supporting surface to the top of the platform shall not be more than 1.5 m to 2.2 m in un-laden condition.

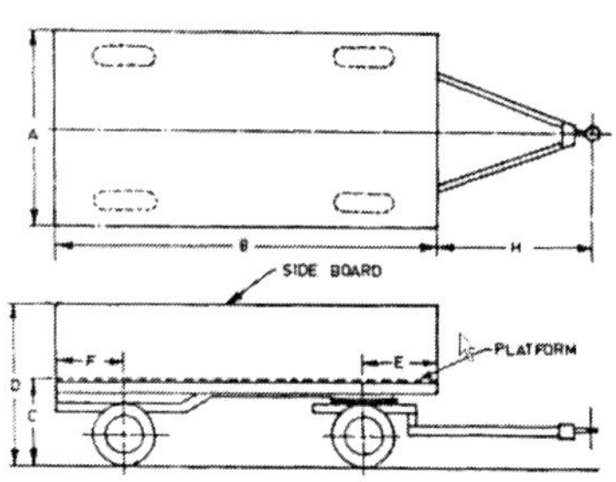

Balanced Trailer

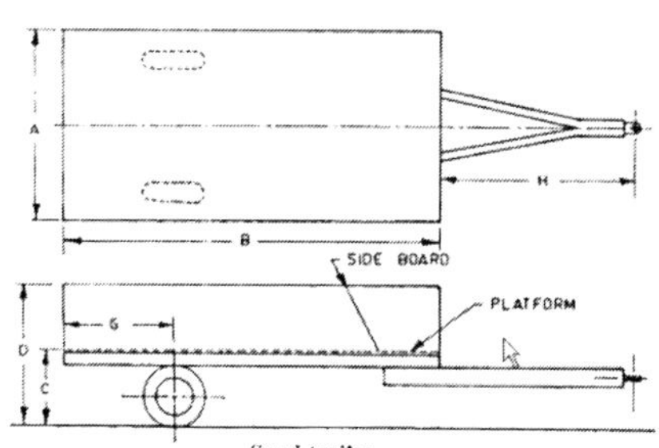

Semi-trailer

❖ In balanced trailer, the distance from the edge of the platform to the centre of front wheel and from rear edge of the platform to the centre of rear wheel shall not be more than 0.8 m.

❖ If the trailer is loaded uniformly with 25% extra pay load and operated at average speed of 25 km/h to 30 km/h for 50 hours will not show any sign of breakage or deformation in any of its part.

❖ The trailer carries lashing hooks for tying down the load.

❖ In case of single-axle trailer, tow eye of drawbar or hitch of the trailer, when fully loaded may be parallel to the level surface after hitching with towing tractor, for balanced trailer drawbar or hitch of the trailer that can be hinged type, as load from trailer is not transferred to the towing tractor.

❖ In towing hitch, a suitable shock absorbing device is present.

❖ The hitch height of trailer is always below the rear axle height of the tractor.

❖ Tow eye of trailer is capable of rotating at 360° angle to take on even rigid trailer hook of the towing tractor.

❖ In balanced trailers, the front axle have capability to swivel to maximum of 120° about the vertical axis.

❖ The trailer has pneumatic tyres and rims which can be amended from time to time.

❖ The axles in trailer conform to the need and can be amended from time to time.

❖ The trailer needs to have suitable spring suspension for recommended load carrying capacity.

❖ In case of tipping type trailer, there is need for tilting the platform without disturbing the trailer.

❖ It carries a single telescopic tipping cylinder.

❖ In this, the hydraulic ram of suitable capacity compatible with tractor hydraulic should be there in case of any tipping.

❖ The ram in this is used in single or tandem as required with operating pressure not more than 13.8 MPa.

❖ Tipping angle of the body in case of tipping type will be around 42° to 50°.

5.3.4 Trailer Weight

Rigid Drawbar Single Axle Agricultural Trailers:

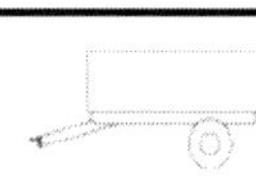

Axle spacing	National Weight Limit Hook/Pin & Eye Type Coupling	National Weight Limit 80mm Ball & Spoon Type Coupling	Technical Requirements
N/A	13 tonnes (10 tonnes on axle & 3 on drawbar)	14 tonnes (10 tonnes on axle & 4 on drawbar)	No Suspension Required

Rigid Drawbar Tandem Axle Agricultural Trailers:

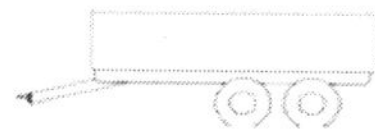

Axle spacing	National Weight Limit Hook/Pin & Eye Type Coupling	National Weight Limit 80mm Ball & Spoon Type Coupling	Technical Requirements
Less than 1.3 metres	**19 tonnes** (8 tonnes per axle & 3 on drawbar)	**20 tonnes** (8 tonnes per axle & 4 on drawbar)	No Suspension Required
1.3 metres or greater	**21 tonnes** (9 tonnes per axle & 3 on drawbar)	**22 tonnes** (9 tonnes per axle & 4 on drawbar)	Flexible Suspension Required
1.3 metres or greater	**23 tonnes** (10 tonnes per axle & 3 on drawbar)	**24 tonnes** (10 tonnes per axle & 4 on drawbar)	Flexible Suspension & Flotation Tyres (≥500mm wide) Required
1.8 metres or greater	**23 tonnes** (10 tonnes per axle & 3 on drawbar)	**24 tonnes** (10 tonnes per axle & 4 on drawbar)	Flexible Suspension & Steering Axle Required

Rigid Drawbar Triaxle Agricultural Trailers:

Axle spacing	National Weight Limit Hook/Pin & Eye Type Coupling	National Weight Limit 80mm Ball & Spoon Type Coupling	Technical Requirements
Less than 1.3 metres	**24 tonnes** (7 tonnes per axle & 3 on drawbar)	**25 tonnes** (7 tonnes per axle & 4 on drawbar)	No Suspension Required
1.3 metres or greater	**27 tonnes** (8 tonnes per axle & 3 on drawbar)	**28 tonnes** (8 tonnes per axle & 4 on drawbar)	Flexible Suspension Required
1.3 metres or greater	**30 tonnes** (9 tonnes per axle & 3 on drawbar)	**31 tonnes** (9 tonnes per axle & 4 on drawbar)	Flexible Suspension & Flotation Tyres (≥500mm wide) Required
1.8 metres or greater	**33 tonnes** (10 tonnes per axle & 3 on drawbar)	**34 tonnes** (10 tonnes per axle & 4 on drawbar)	Flexible Suspension, Flotation tyres (>500mm wide) and Steering Axles (i.e. Both Foremost & Rearmost Axles) Required

Drawbar/Centre Axle Agricultural Trailers:

	Axle spacing (X)	National Weight Limit
Double axle 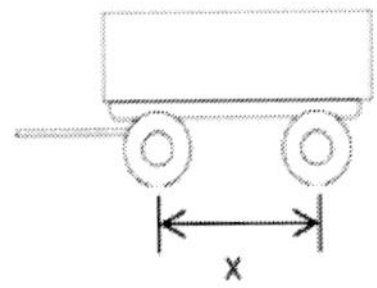	Less than 3.0 metres	**16 tonnes** (8 tonnes per axle)
	3.0 metres or greater	**18 tonnes** (9 tonnes per axle)
Triaxle 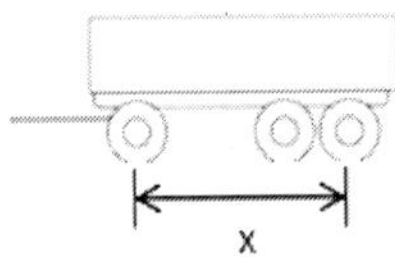	**Tonnes/Metre (X)** 5.5 tonnes	**25 tonnes**
Four axle 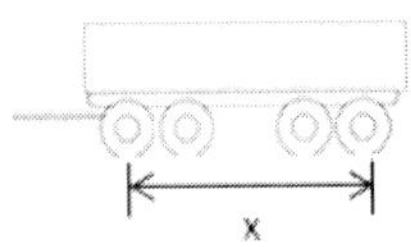	**Tonnes/Metre (X)** 5 tonnes	**30 tonnes**

5.4 Classify CAT I and II implements

- Agriculture implements are categorized as 1 and 2 based on the dimensions of three point hitch pyramid.
- The three-point hitch is often refers to as the way the ploughs and other implements are attached to the tractor.
- The three points resemble either a triangle or the letter A.
- The essential dimensions of hitch pyramid are given in table.

Table: Essential dimensions of three point linkage

Category	Hitch pin size, mm		Lower hitch point span, mm (L)	Mast height, mm (h)	PTO power at rated engine speed, kW
	upper hitch pin hole dia (d1)	lower hitch pin dia (D2)			
1	19.3	22	683	460	Up to 48
2	25.7	28	825	610	Up to 92

5.5 Select proper implement for a specific function

- Tractor drawn implements can be either of trailed, semi-mounted or mounted type.
- A trailed implement is that which is attached to the tractor drawbar by a pin joint.
- A semi-mounted implement is that which is rigidly attached to the tractor and has a wheel or wheels to support part of its weight.
- A mounted implement is that which is attached to the tractor as an integral part which is hydraulically controlled and is kept raised during transport.

5.5.1 Selection

The selection of tractor and its matching implements is quiet difficult due to availability of variety of tractor/implement models.

Factors to be considered while selecting an implement:

- specifications of tractors and implements
- tractor/implement performance data
- soil and operating conditions
- nature of work to be carried out
- tillage practices
- crop mix
- number of crop acres
- field capacity

The tractor implements can be classified as:

- agricultural or landscape
- tillage, maintenance or materials handling
- front, mid or rear

Matching the tractor–implement sizes is difficult due to:

- diversity in the makes, types and sizes of tractors and implements
- variability of weather conditions that constraint the available working days
- temporal and spatial variability of soil due to differences
- due to soil tractive force, moisture content and soil texture
- financial and administrative problems of funding
- availability of spare parts
- maintenance facilities
- labour constraints
- diversity of cropping patterns and intensities

5.6 Take precautions before using Implements and Trailer

Tractor operator should have an understanding about the function, operation and limitations of the implement which he/she is operating. Before using the implement and tractor, an operator should take certain precautions before using the implement and tractor.

5.6.1 Precautions in using implements

- In using plough or other implements, operator should note that the implement should be properly hitched with the tractor.
- The implement should be set and adjusted properly.
- In the case of tractor drawn ploughs, adjustments are done on rear furrow wheels. When these are properly adjusted, there should be about 6 to 12 cm clearance between the rear side of the landside of the back plough bottom and furrow wall.
- Before using implements, the operator should check for proper adjustment, cutting edges, repair of broken parts, replacement or repairs of worn out parts and should replace the missing parts such as nuts, bolts, pins grease nipples etc.

Using Towing Implements

Before using the towing implements, the tractor operator should:

- fit the attachments as per manufacturer's instructions.
- attach the implements to the draw bar or mounting points as provided by the manufacturer.
- not change, modify or raise the height of the draw bar unless provided for by the manufacturer.
- regularly check the safety pins on towed lift-wing implements, to ensure they are not worn.
- ensure that all guards on towed implements are in place.
- note that the hitch should not be above the centerline of rear axle, around the axle housing or to the top link pin.
- never attach implements unless the PTO shaft is guarded.

5.6.2 Precautions before using trailers

Before using the trailer, the tractor operator should:

- see that proper and approved trailer should be used.
- check for correct tyre pressure of tractor and trailer wheels.
- properly attach the trailer with the tractor using approved hitch mechanism.
- not keep the drawbar of trailer above the center line of rear axle.
- use proper ballast on tractor front/rear to arrest front end lifting to reduce wheel slip.
- keep trailer brake in good condition.
- keep all electrical lights on tractor and trailer in working condition.

- avoid excessive loading.
- avoid frequent braking.
- always connect wheeled trailers to the drawbar and check for safety chain or cable.
- use correct high-strength steel hitching pin designed for the tractor and the task.

- not repair or weld coupling pins and towing hooks.
- securely retain coupling pins by locking mechanism.
- not repair tow-eyes, but should replace them if they are worn beyond 10%.
- fit a safety chain between a tractor and towed trailers or implements, except those carried on a three-point linkage.

Chapter 6

Land Preparation Requirement & Land Opening

6.1 Introduction

- Land preparation provides necessary soil conditions which will enhance the successful establishment of young offshoots or tissue culture plants which are received from the nursery.
- Well-prepared land, control the weeds, recycles plant nutrients, and provides a soft soil mass for transplanting and suitable soil surface for direct seeding.
- Land preparation covers a wide range of practices from zero-tillage or minimum tillage which minimizes soil disturbance through to fully puddled soil that actually destroys soil structure.

6.1.1 Factors considered during land preparation and opening

- Availability and quality of irrigation water;
- Field selection;
- Mechanical actions to be implemented;
- Chemical needs for pre-plant soil improvement;
- Tools and equipment needed for date cultivation;
- Labour needs;
- Irrigation design and installation;
- Leaching schedule;
- Hole preparation;
- Financial requirements
- Time schedule.

6.2 Preparation Implements

6.2.1 Primary land preparation

Mould Board Plough

- It is an important plough for primary tillage in canal irrigated or heavy rain areas where too much weeds grow.
- To completely invert and pulverize the soil, up-root all weeds, trash and crop residues and bury them under the soil.
- It is used for turning and mixing green manure crop for decaying under the soil, which adds humus to the soil.

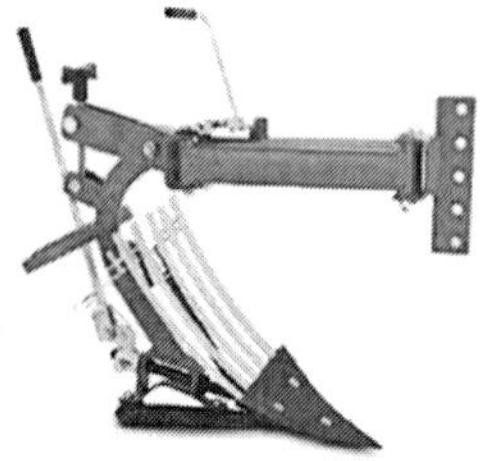
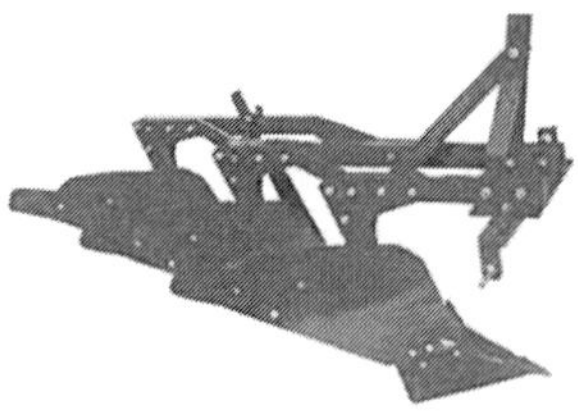

Disc Plough

- It is useful in hard and dry, trashy, stony or stumpy land conditions and in soil where scouring is a main problem.
- It is suitable for land where there is much fibrous growth of weeds as the disc cuts and incorporates the weeds.

Secondary tillage implements

- Lighter or finer operations done on the soil after primary tillage is known as secondary tillage.
- After ploughing, the fields are left with large clods with some weeds and partially uprooted stubbles.
- Harrowing is to be done to a shallow depth to crush the clods and to uproot the remaining weeds and stubbles.

Rotary tiller

- ❖ The rotary tiller is an important implement which gives fine degree of soil pulverization.
- ❖ It is directly mounted to the tractor and operated.

6.3. Field Management

6.3.1 Land preparation

For preparing land:

- ❖ Plough the land during summers to economize the water requirement for initial preparation of land.
- ❖ Fill the fields with water for 1 or 2 days before ploughing and allow the water to soak.
- ❖ Keep water to a depth of 2.5cm at the time of puddling.

Special process for difficult soils:

- ❖ In case of fluffy paddy soil, compact the soil with heavy stone roller or oil-drum with stones inside once in three years to prevent sinking of draught animals and workers during puddling.
- ❖ In case of sodic soils with pH more than 8.5, plough the land at optimum moisture regime by using gypsum uniformly and impound the water and provide drainage for leaching of soluble salts and finally use green leaf manure.
- ❖ In case of saline soils with EC values above 4 dS/m, provide lateral and main drainage channels, apply green leaf manure with 25% extra dose of nitrogen along with P and K and $ZnSO_4$ at planting.
- ❖ In case of acid soils, use lime based on the soil.

6.3.2 Field Preparation for Irrigated Crop

Ploughing

- ❖ Plough the field with iron plough once or twice.
- ❖ Sorghum does not require fine tilth as it adversely affects germination and yield in case of direct sown crop.
- ❖ To overcome subsoil hard pan in Alfisols, chisel the field after every 0.5 m intervals to a depth of 40 cm on both the sides of the field.

- ❖ Use disc ploughing once and cultivator ploughing twice in order to increase the yield of sorghum and the succeeding blackgram.
- ❖ Use FYM and 100% N as specified.
- ❖ In soils with sub-soil hard pan, chisel should be done every year at the start of the cropping sequence to create a favourable physical environment.

Formation of Ridges and Furrows

- ❖ Form ridges and furrows using a ridger at 6 m long and 45 cm apart.
- ❖ Form irrigation channels across the furrows.
- ❖ Form beds of size 10 m^2 and 20 m^2 depending on the availability of water.

6.4 Select and suggest the implements as per the need of the farmer

6.4.1 Primary Land Preparatory Tillage Implements

Deep tillage implements are needed for turning under the soil to properly bury the incorporated crop residue. For doing this, the ideally suited implements are:

- ❖ Mould board ploughs
- ❖ Soil inverting ploughs
- ❖ Deep disc ploughs
- ❖ Off-set disc harrow
- ❖ Tine cultivators
- ❖ Rotavators

6.4.2 Secondary Land Preparatory Tillage Implements

- ❖ Off-set disc harrows
- ❖ Cultivators
- ❖ Gang type harrows

- ❖ Spring harrows
- ❖ Tooth tyne harrows
- ❖ Plankers
- ❖ Rollers
- ❖ Puddlers
- ❖ Land planers
- ❖ Wheel cages & weights
- ❖ Furrow openers & bund farmers

Apart from these, there are certain commonly used hand tools that are used by farmers for preparing their land, which includes:

- ❖ Spade (for bund making, digging channels & other sundry works)
- ❖ Jandra (hand-operated /manual bund maker-cum-soil mover/land leveller)
- ❖ Sickle (for harvesting- serrated toothed sickles
- ❖ Spray-pumps (for insecticide-pesticides application; nutrient sprays or herbicide spray)
- ❖ Dusters (for insecticide-pesticides application ; nutrient sprays or herbicide spray)
- ❖ Fork (for lifting/gathering crop residue).

6.5 Calibrate seed drill as per requirement

6.5.1 Seed drill

It is a device which sows the seeds for crops by metering out each seeds, positioning them in soil and covering seeds to certain depth.

It allow farmers to sow the seeds in well spaced rows at particular depths at particular seed rate where each tube creates a hole of specific depth thereby dropping seeds and covering them.

It comprises of hopper, filled with seeds which are arranged above series of tubes that can be set at required distances from each other for optimum growth of resulting plants.

Pre Operation:

Before using seed drill:

- ❖ ground should be plowed and harrowed
- ❖ harrow should smooth the soil and breaks any clumps
- ❖ set the drill as per size of the seed used
- ❖ put the grain in the hopper on top, following from behind as per spaces and seed

Working

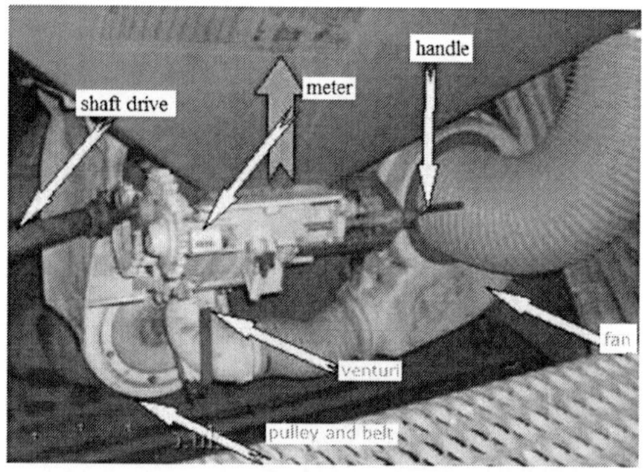

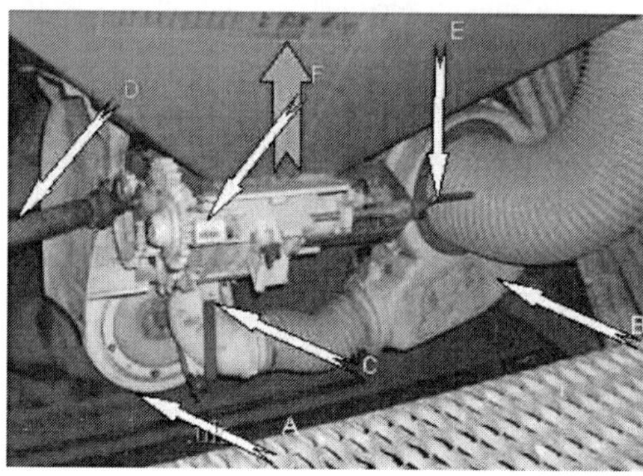

- In seed drill, seed metering unit is kept at the base of seed hopper where the fan (B) is driven from rear of power harrow through pulley and belt (A).
- The airflow is forced through a venturi at (C) where seed gets dropped to airflow.
- The seed/air mixture moves up a tube inside the seed hopper in direction of arrow shown.
- The seed rate is maintained at a rate which should be proportional to ground speed by shaft drive (D).

- ❖ Adjustment of the rate is made by handle (E) which opens or closes the seed aperture.
- ❖ The (F) is a meter that will record the area in hectares which are sown.
- ❖ Air and seed in this, moves by the tube (G) towards distribution head (H) which is located in seed hopper, where it swirls, till it passes to one of the tubes (J) which leads to coulters.
- ❖ The solenoid valves (K), shut off the grain flow during the tramline.

- ❖ In this, the wheel (M) drives the metering unit by the shaft (N).
- ❖ The hoses (L) carries the seed to the coulters (O) and the seed passes down the tube of the coulter (P).

- ❖ Here, the ceramic tip, cuts the groove in the soil for the seed that exits at (R) for fall.
- ❖ The pressure on the coulter gets adjusted by tensioning the spring (S).
- ❖ Plastic plug (T) swings in the seed aperture which saves blocking when drill reversed whilst in the soil.

6.5.2 Seed Drill Calibration

Seed drill calibration involves following steps:

Step 1: Determine area seeded when driving calibration distance

- Calibrate the drill for a driving distance of 200 feet which shows an arbitrary length.
- Determine the width seeded by the drill.
- Calculate the working width by multiplying the space between seed openers by number of openers on the drill.

Step 2: Determine the number of revolutions of grain drill drive wheel when traveling the calibration distance

- Use the measuring tape or other measuring device to measure a course which is 200 feet in length.
- Determine which wheel is responsible for driving the seeding mechanism on the drill.
- On some drills it is a special wheel which only runs the seed metering mechanism, while on other drills it is one of the wheels that supports the drill.
- Determine how many revolutions the wheel makes in the field in a 200-foot length.
- Now make a mark or apply a piece of heavy tape to the wheel to count the wheel revolutions while driving.
- After driving 200 feet, count the number of revolutions made during the 200 feet.
- For more accuracy, repeat the step one or two times and take the average.

Step 3: Set up the drill for calibration

- Fixed calibration is done by simulating drill operation.
- The seed delivery system of the drill is operated much in the same way as it is when being pulled across the field by turning the drive wheel.
- For stationary calibration, lift the drive wheel above the soil surface so that it can freely spin.
- Keep the drill securely attached to the tractor to maintain a level of safety when elevating the drive wheel and to operate the hydraulics on the drill.
- It is required to lower the drill as it permits the drive mechanism to get engaged so that seeding rate calibration can be accomplished.

- Use rate chart given by drill manufacturer and adjust the seed meter for the desired seeding rate of the species of interest.
- If the seed is not on the chart, then estimate a seeding rate by selecting a similar seed size, so that you should have an approximation.
- Make sure that required seed has been added so that seed cups are full and seed metering fluted wheels or sponges are completely covered for calibration process.
- Collect the seed by using large tarp by placing under all drill openers so that seed falling through the seed delivery tubes and disk openers can be collected from below the seed delivery mechanism and weighed.
- The seed can also be collected by removing the seed tubes from the disk opener assembly and placing them into several buckets or other containers for catching and transferring seeds to one container for weighing.
- Now start collecting seed for calibration by ensuring that the drive wheel has been turned to a convenient starting point and that the tarp or buckets are clean.

Step 4: Calibrate the drill

- Adjust the drive wheel for number of revolutions required to show 200 feet of movement.
- Collect all the seed from the tarp or buckets and weigh them.
- Use a balance with a gram scale and convert grams to pounds.
- Collect the seed from simulated seeding of fraction of an acre, weighed and compare the desired seeding rate.
- If the seedling rate is too low, the metering mechanism is adjusted to deliver more seed and the process is repeated.
- It is required to maintain a record of seeding rates for each setting.
- Determine the seeding rate of the drill.
- For checking each opener for consistency, collect the seed from individual seed tubes and record the weight of the seed from each tube.

6.6 Take up the following function:

6.6.1 Levelling of field

Land levelling is a measure which is used in surface irrigation like basin and furrow irrigation which comprises of:

- preparing the irrigation plot such that no high/low spots disturbs the uniform distribution of irrigation water on the field
- ensuring optimal slope for water movement across a field when irrigated.

Levelling lead to efficient irrigation and when fertigation and chemigation are used, in more efficient use of fertilizers and pesticides. In an unlevelled field, high spots might not be covered by irrigation water, and the dissolved nutrients and/or pesticides can percolate unused into the soil. In case of low spots, water and the dissolved nutrients and chemicals might accumulate there and create zones of water logging and nutrient or pesticide accumulation. This will disturb the soil aeration and water uptake by crops.

Levelling of land can be done manually or with machinery with required equipment. Heavy earth movement should be avoided for keeping natural soil structure undisturbed and in maintaining good growing conditions for crop roots and keeping costs low. Experienced farmers do land levelling as per visual assessment like:

- Use of hoes on small plots
- Using draft animals and equipment

Advanced and sophisticated levelling equipment is operated with laser emitter, a laser sensor or receiver, and scraper which are pulled by tractor. After the desired level or slope of the field and/or the difference of the high and low spots have been surveyed, the emitter is set to send a rotating laser beam creating a plane of laser light above the field surface. The laser light is used as levelling reference that directs the hydraulic system of moving tractor and scraper which controls the levelling.

6.6.2 Intercultural operation (Weed control and tilling)

Weed control

Weed control involves strategies that prevail weed species with their:

- size,
- condition
- age
- soil condition
- equipment
- crop plant size
- weather

There are three approaches for controlling weed such as:

- mechanical
- chemical
- physical

Mechanical weed control

- It involves mowing or cultivating around planted seedlings.
- Mowing and tilling are beneficial to plantation establishment, increasing growth, survival and nutrient uptake.
- The effects of mechanical weed controls are short-lived and may need to be repeated five to six times per growing season.
- It is very effective for removing existing vegetation.
- It creates favorable growth conditions for existing seed beds of undesired vegetation.
- It increases future growth of competition.

Chemical weed control

- It is an efficient method of weed control, but requires knowledge and skill to use safely and effectively.
- In this, herbicide use and registration is controlled by local agencies.
- It needs planning and preparation.
- In this, the land manager must consider the species to be controlled along with crop trees.
- The herbicides are designed to eliminate undesirable or problematic species through various modes of action.

Physical weed control

- In this, physical barriers or mulching systems involves installation of manufactured or organic barriers to control vegetative competition and enhance microclimatic conditions for planted seedlings.
- Such system is not a common means of weed control due to high upfront costs and labor intensive installation.
- It is more practical for small plantings.

Tillage

Organic mechanical weed control has four distinct phases, each one very important to the overall success of weed control program which involves:

- Tillage
- Planting
- Blind cultivation
- Row cultivation

The idea of early mechanical weed control is to remove the bulk of weed population before it competes with crop and to create as large a crop-to-weed size differential as early as possible. In this, when crop plants are bigger and more vigorous than weeds, the weed pressure will usually not jeopardize the crop, so effective early weed control is required before weeds present a visible threat to the crop.

Appropriate tillage of fields is important to:

- Create a good seed bed for uniform, vigorous crop emergence
- Prepare ground adequately for successful subsequent mechanical weed control operations
- Kill weeds that have already emerged

6.6.3 Puddling

- It is a tillage system in which the soil is repeatedly plowed and harrowed under submerged conditions, to make it soft for transplanting and less permeable to water.

- ❖ In this, many tillage implements are used to make the soil impermeable by manipulating it in standing water that reduces apparent specific volume and facilitate transplanting.
- ❖ In this, cultivation requirements are mixing and incorporation of organic matter, compaction of pan by smearing and hydraulic sealing, without increasing pan-depth.
- ❖ The tillage implements used for puddling are backward inclined narrow tines, discs, powered rotary cultivators, puddlers, conical rotor, etc.
- ❖ It is achieved by repeated intensive tillage under ponded-water conditions that serves to break down soil aggregates, reduce macro-porosity, reduce soil strength in the puddled layer, disperse fine clay particles, and form a dense zone of compaction in subsoil.

6.6.4 Side dressing of fertilisers in row crops

- ❖ It refers to those fertilizer materials that are applied to the soil, during the early to mid-growth period of the crop.
- ❖ The side dressed fertilizer materials can be either broadcast on both sides of the row about 6 to 8 inches from the plants or banded down the row middles.
- ❖ Nitrogen fertilizers are more often side dressed than phosphorus or potassium materials which serves as supplement nitrogen needs and ensures more efficient use under cropping conditions favoring nitrogen loss.

6.6.5 Spraying pesticides and herbicides

Many farmers choose to use chemicals to keep weeds and pests from destroying their crops and to add more nutrients to the soil. There are three different kinds of pesticides; herbicides, insecticides and fungicides. All three of these pesticides are used to kill different kinds of pests that can be found on a farm.

Pesticides and herbicides can be sprayed using:

- ❖ Hand sprayers and atomizers
- ❖ Hand compressed sprayers
- ❖ Knapsack sprayers
- ❖ Tractor-mounted sprayer
- ❖ Motorized knapsack mist blowers
- ❖ Ultra low volume or controlled-droplet applicators

Before selecting the sprayer to use, check:

- ❖ label and IPM guidelines for particular pesticide that will be applied.
- ❖ Recommended application rate
- ❖ type of equipment best suited for application,
- ❖ style of nozzle and information for best results
- ❖ sprayer for particular crop type

- for sprayer droplet size and coverage

Spraying technique

Technique to spray the crop is important in order to have good coverage which should include:

- fitting the type of equipment that is to be used.
- sweeping motion over foliage by allowing spray material to penetrate in leaves using hand-held gun.
- requirement of airflow pattern for all plant canopy to receive pesticide while using fixed-fan type units
- enhancement the movement and distribution of fog and mist particles.
- proper use of sprayer with required coverage.
- to monitor the coverage with hydraulic or mist sprayer using strips of water-sensitive paper.
- evaluation of coverage using fluorescent dye in water.

6.6.6 Harvesting of crop with suitable machinery attachment

It is the operation of cutting, picking, plucking digging or a combination of these operations for removing the crop from under the ground or above the ground and removing the useful part of fruits from plants.

Harvesting action can be done by four ways

- Slicing action with a sharp tool.
- Tearing action with a rough serrated edge.
- High velocity single element impact with sharp or dull edge.
- Two elements scissors type action.

Harvesting can be done by

- Manually operated tool
- Animal drawn machine
- Mechanically operated machine

Harvesting machinery

- Combine Harvester: It is used for harvesting many grains and seed crops.
- Self-Propelled Riding Type Vertical Conveyor Reaper: The riding type vertical conveyor reaper is a self-propelled unit used for harvesting rice, wheat, soybean and other cereals and oilseed crops.
- Tractor Mounted Vertical Conveyor Reaper: Vertical conveyor reaper is used for harvesting and windrowing of wheat and paddy crops.
- Tractor Mounted Combine Harvester: Combines are used for cutting, threshing and cleaning of cereal and other crops in one operation.

- ❖ Self-Propelled Combine Harvester: Self-Propelled Combines are used for cutting, threshing and cleaning of cereal and other crops in one operation.
- ❖ Self-Propelled Reaper Binder: Reaper-Binder is a unique harvesting machine that reaps the crop as well as binds it simultaneously.
- ❖ Groundnut Digger Shaker/Harvester: Groundnut digger shaker are used for harvesting groundnut crop.
- ❖ Sugarcane Harvester: It is used for harvesting sugarcane. It de-tops the cane from the top, cuts it from the base, strips the cane, cuts it into billets, removes the trash and simultaneously loads the cut billet into the trolley moving by its side.

6.6.7 Adjustment of Top link

- ❖ A top link is the upper centre arm of a tractor's three-point hitch which is used to attach implements to the tractor such as ploughs and hay tines.

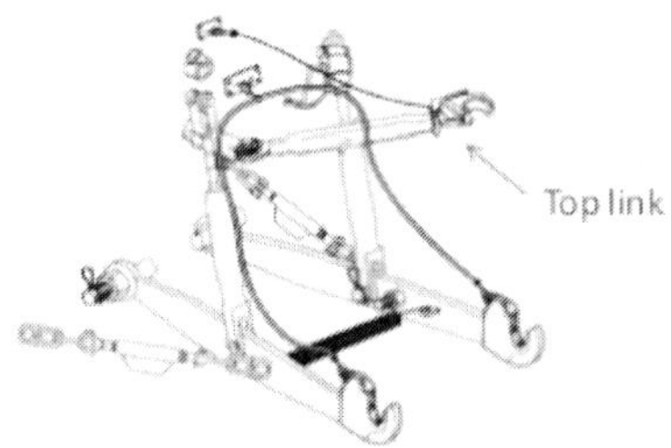

- ❖ Top links have two ends, one end attach to the tractor while other end attaches to the implement.
- ❖ The dimension of a top link end has several categories such as "Cat 0" with end length 16mm, "Cat 1" with end length 19mm and so on.

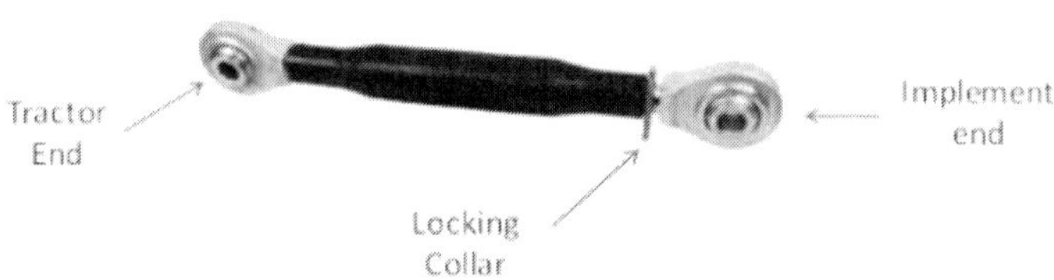

There are two types of top links:

1. Manual
2. Hydraulic

To select the correct top link to do the job, it is required to check for:

- ❖ particular top link working with required tractor make.
- ❖ the standard or heavy duty top link.
- ❖ type of ball end that fits on to the tractor as required.
- ❖ ball end which attaches to the implement parts which is selected.

Adjustments:

For plough and other implements to work correctly, there is a need of adjustment of top link. For adjusting the top link follow the procedure:

- ❖ select the implement which needs to be connected with the tractor by top link

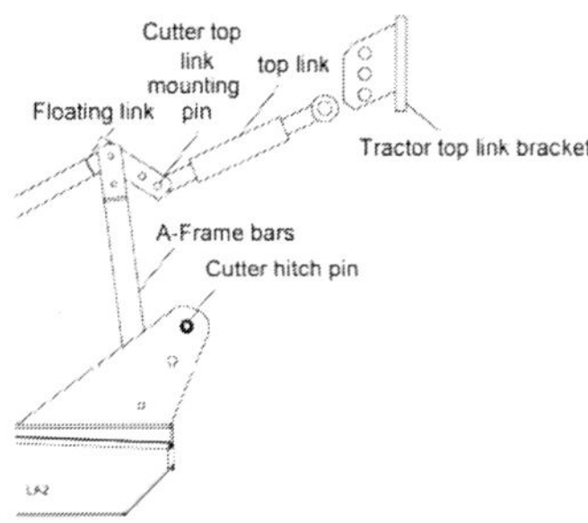

- ❖ measure the exact distance of the tractor and the implement.
- ❖ check the length of top link.
- ❖ keep the top link holes on the tractor aligned with the holes on the implement
- ❖ keep the arms just above the implement arm pins till the left arm ball end is lined up
- ❖ attach the top link to lowest hole provided in tractor's top link bracket.
- ❖ adjust the top link length in such a way that it should freely move up and down, left and right without any obstruction
- ❖ attach the rear portion of tractor top link to the first hole on the cutter floating link.
- ❖ select a top link mounting pin that will allow the floating link to swing freely through the cutter A-frame bars.
- ❖ raise cutter to transport position and adjust tractor top link until cutter is level in the raised position.
- ❖ once the implement end is fixed, lock it using the collar

6.6.8 Adjustment of Chain guider

It is imperative that the correct adjustment be maintained. Only through correct adjustment can a smooth and level finish be achieved in field working.

Loose chains will lead to:

- ❖ Uneven performance across the width of the machine
- ❖ Uneven weed control
- ❖ Unsatisfactory incorporation
- ❖ Ineffective levelling
- ❖ Accelerated or premature chain wear
- ❖ Chains failing to engage with transport locators when folded
- ❖ Machine damage when folding or unfolding

❖ Uneven field surface with ridges and furrows being created.

Chain mount arm height adjustment

❖ Use tractor hydraulics to raise the chain mount plate for adjustment.

❖ Place a suitable block or stand to support the arm.

❖ Using tractor hydraulics, lower the machine till the height adjusting chains are loose.

❖ Remove spring retaining clip.

❖ Slide chain links through slot in lifting arm and turn links to 90 degrees for fine adjustment.

❖ Install the retaining clip.

❖ Remove the stand.

6.6.9 PTO speed as per selected PTO operated implements, equipments and machinery

❖ PTO and its associated shafts and universal joints are common cause of incidents and injury in farming.

❖ Power Take-Off shaft is an efficient way of transferring mechanical power between farm tractors and implements.

❖ A PTO shaft rotates at 540 rpm or 1,000 rpm of speed.

❖ The dual speed PTO's have 630 to 1,170 rpm speed.

6.6.10 Measurement of wheel slippage

- Tractor tires slip when they turns faster than the ground speed.
- Less than 60% to 70% of power developed by tractor engine is used to pull an implement across the soil.
- Slip provides safety valve against shock overloads which damages the power train.
- Wheel slippage is required to reduce wear on the power train.
- If the slip is very less, it will anchor the tires to the soil due to less ballast.
- Wheel slip for 2WD tractors using tyres should be between 10 and 15 percent.
- Wheel slip for 4WD/front wheel assist tractors with tyres, will be in between 8 and 12 percent.

Measuring Wheel Slip

To determine the percentage wheel slip:

- Place a mark on the inside of the tire which can be observed from tractor seat.
- Mark off a distance of 100 feet part way into the field.
- Now find the circumference of the tire by placing a string/measuring tape around the center part of the ribs.
- Now drive the 100 feet with the field operation in progress, by counting number of revolutions of the tire.
- Multiply the number of revolutions by the circumference (in feet) of the tire and then divide it by 100.

Adjusting Wheel Slip

If excessive slippage more than 15% occurs then:

- add weights or fluid in tires
- change the air pressure
- add duals
- purchase new tires, if tyres are heavily worn

6.6.11 Fuel Consumption/hr or acre

- The tractor of 55 bhp consumed maximum fuel on hourly basis at all the three load settings, due to higher horse power.
- In case of unplough soil at 2/3rd load setting, fuel consumption by 35 and 30 bhp were 3.2 I/h and 2.9 I/h for cultivator and 3.2 I/h and 2.7 I/h for disc harrow.
- Specific fuel consumption for cultivator was 88.53, 68.77 glbhp-h and, for disc harrow was 88.53, 64.03 glbhp-h.
- As noted, specific fuel consumption decreased with increased in tractors horsepower.
- In stubble soil at full load condition, fuel consumption on hourly basis for 55, 35 and 30 bhp were 4.3, 3.8 and 4.0 I/h for cultivator and 3.5, 3.4 and 3.8 I/h for disc harrow.

- Specific fuel consumption in unplough condition were 55.84, 90.17 and 11.067 glbhp-h, for disc harrow and 72.44, 101.97 and 127.27 glbhp-h for cultivators.
- In tilled irrigated condition at full load, fuel consumption for 55,35 and 30 bhp was 5.0,4.6 and 4.9 I/h for cultivator and 4.0, 4.0 and 4.1 I/h for disc harrow.
- Fuel consumption was minimum at 1/3rd part load for all the three tractor implement systems.
- The same increased with increase in load.
- Although, tractor gave the best work performance at full load but, field capacity at 2/3rd and full load setting was not commensurate with increase in fuel consumption at these loads setting.
- At 2/3rd part load setting, tractors gave optimum work performance and efficient fuel consumption.

Chapter 7

Use of Implement for Tractor

7.1 Do Lateral & horizontal adjustments of Implements

There are many attachments and accessories which, though they are not essential to the operation of the tractor, increases its range of usefulness. Some of these improve the efficiency of the engine or the hitching of implements or the grip of the tractor on the soil while some make the tractor more versatile by adding to the comfort of operator, making long hours of work possible.

Some tractors have only fixed drawbar, with no range of hitching height, and only a small range of lateral adjustment which can cause much extra work when various implements of different height have to be used and when it is required to offset an implement sufficiently to be clear of a standing crop or to work near to a hedge.

7.2 Do Hitching and unhitching the implement from tractor

Hitching and unhitching tractors and implements are daily tasks of an operator or workers who are working on the farms. For doing such task, the need for assistance particularly when the tractors or implement is big, as element of risk involved is much more.

The hitch mechanism involving in hitches or unhitches of an implement from a tractor includes making or breaking of hydraulic/electrical connections, which needs to be performed with thorough quality.

Hitching a drawbar:

- Place the tractor to align the hole in the drawbar with the hole in implement hitch.
- Stop the engine, and apply tractor brakes.
- Attach the implement using the proper-sized hitch pin and security clip.
- Raise the implement jack stand and remove chock blocks from the wheels.
- Connect the PTO shaft, hydraulic hoses, and/or electrical connections as required. Refer to the appropriate task sheets on these subjects.

Hitching a 3-point hitch attachment:

- ❖ Move the stationary tractor drawbar forward for clearance.
- ❖ Place the tractor in such a way that pin holes of the draft arms are closely aligned with the implement hitch points.
- ❖ Raise or lower the draft arms to match the implement hitch points.
- ❖ Stop the engine, securely park the tractor, set the brakes.
- ❖ Attach the left draft arm to the implement hitch point using the proper size hitch pin and security clip.
- ❖ After attaching the left arm, connect the right arm.
- ❖ Remount and start the tractor to use the hydraulic system to raise the lift arms if required.
- ❖ Match the top link of the 3-point hitch to the implement's upper hitch point. It can be done by raising the lift arms to lengthen or driving ahead with implement down, to shorten to adjust. It is noted that the implement is not level if the upper link has been adjusted too many times and when it is out of level, in such case, the machine will not work properly
- ❖ Securely attach the upper hitch pin with the proper size hitch pin and security clip.

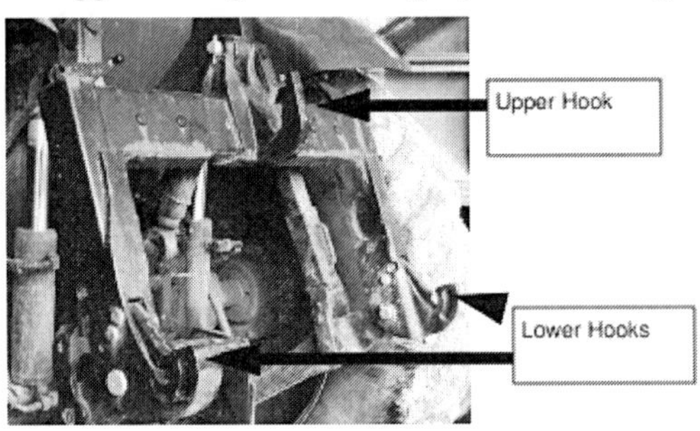

Safety precautions in using 3-Point hitch implement

Following are points, which tractor operator needs to be followed to safely operate the implements, which includes:

- ❖ Take the idea about the size and dimension of an implement in comparison with the tractor.
- ❖ Make sure that the machine is in the transport or up position for travel on the fields.
- ❖ Lower the machine to the field position when ready to use it, as it will keep the load pulling below the center of gravity.
- ❖ Engage the machine operation mechanism (levers, pins, etc) for the wheels to power the machine if using ground-driven machine.
- ❖ Start field operation of the machine by giving attention to field boundary fences and obstacles.

- ❖ Allow space at ends of rows or fields to lift the equipment with the 3-point hydraulic lift.
- ❖ Do not make turns with a 3-point hitch implement in or on the ground, as it places the undue force on the 3-point hitch draft and lift arms that causes damage to the machine.
- ❖ Lift the implement to the transport position before using, as ground driven implements which are used on the road can damage the road surface.

Safety guidelines

It is the duty of the tractor operator to check and do following, while hitching or unhitching the implement with the tractor which includes:

- ❖ Reading of the owners manual before use.
- ❖ Make sure the hitch category of the equipment matches with the tractor hitch category.
- ❖ While do hitching and unhitching, make sure that both, tractor and implement should be on level surface.
- ❖ If there is a risk of rolling, block the tractor or implement wheels before unhitching.
- ❖ Hitching of trailed equipment only to the tractor drawbar, as hitching the implement elsewhere will displace the centre of gravity on the tractor which results in backward overturn.
- ❖ Use low gears and low engine RPMs to reduce sudden movements when approaching of pulling away from an implement.
- ❖ While doing hitching/unhitching of implement, keep visual contact and communication with tractor ground assistants all the times.
- ❖ Ground assistants should stay clear of wheels until the tractor and implement are lined up for hitching.
- ❖ Use of only approved hitch pins and should replace the hitch pins that are worn, bent, or broken.
- ❖ Use of properly rated safety chains with pins and balls of proper size, as strength of safety chain must be equal to the gross weight of the load being towed. Be sure no loose chains are dangling either from the drawbar or the implement.
- ❖ Using locking pins on hydraulics.
- ❖ Lock or secure hitch pins with retainer clips.
- ❖ Shut off the engine, place transmission in park and set the parking brake on the tractor before connecting hydraulic lines, electrical lines, and/or PTO shafts.
- ❖ Make sure pressure is released before connecting/disconnecting hydraulic lines.
- ❖ Use proper lifting techniques to reduce sprains/ strains when moving or lifting the implement tongue.
- ❖ Use approved size tongue jack that is attached to the implement to support the implement as temporary jacks can kick out or fail, if the implement moves.

- Remove all connections before moving the tractor away from an implement after unhitching.
- Hitch/unhitch on level sites.
- Make sure all shields and guards are in good condition and are properly installed.

Chapter 8

Tractor Operation

8.1 Functions of gauges

Many tractors are equipped with gauges which can be:

- fuel pressure gauge
- oil pressure gauge
- water temperature gauge
- hour meter
- hydraulic pressure gauge
- temperature gauge

Fuel Pressure Gauge

- It is a good diagnostic tool which keeps an eye on the fuel system of the tractor.
- It ensures peak performance level of fuel pressure.
- It monitors the fuel pressure in an engine which can prevent damage to fuel pump and injectors.
- It helps in diagnosing an overall loss of engine horsepower.

Oil pressure gauge

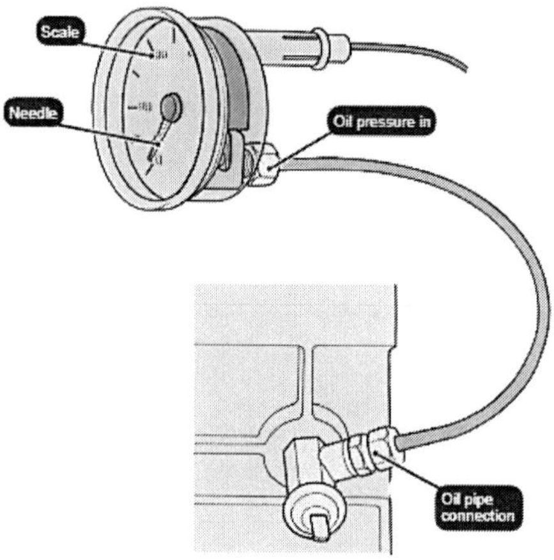

- ❖ It is an important instrument in a tractor which acts as an indicator of the engine's overall state.
- ❖ It gives early warning system and gives advance notice of any problem inside an engine.
- ❖ In this, zero readings with engine running above idle speed shows that the gauge is faulty, oil level is low, broken of oil pump or drive.
- ❖ It monitors the oil pressure in main oil ways which is near the pump and filter by tapping engine block having a sensor or oil pipe take-off.
- ❖ If the gauge is mechanical, it works by literally forcing the needle around the dial under the pressure from the engine.

- ❖ If the gauge is electric, it works by using the sensor screwed into the engine block to produce variable resistance affecting the amount of current that passes through the circuit having gauge and sensor.

Water Temperature Gauge

- ❖ It is mounted on top of the governor housing.
- ❖ It is used to monitor the coolant temperature.

- It allows quickly to determine when the engine fuels have reached their operating temperature.

8.2 Wheel track setting- application & demo

- In tractor, setting of track involves adjustment in tension on tracks.
- If tractor tracks is too loose, it may spool off the undercarriage and in such case it needs replacement.
- If tractor tracks are too tight, it result in damage of undercarriage and will lead to premature wear.

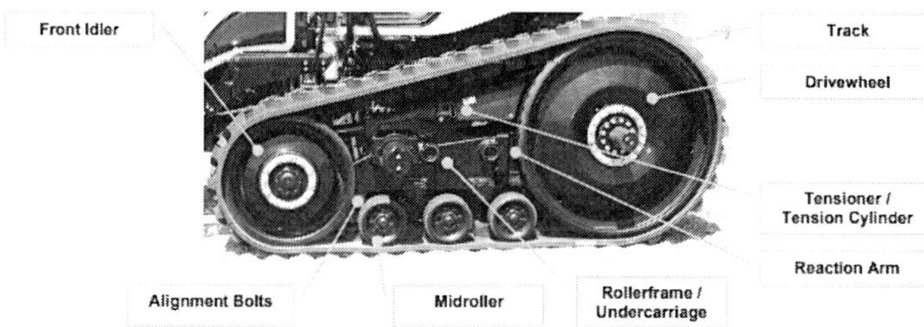

8.2.1 Checking of Track Alignment

- Place the tractor on level ground.
- Keep the transmission in park and stop the engine.
- Place chocks at the front and rear of the tracks to prevent unwanted movement.
- Check the temperatures of the inner and outer face of guide lug using infrared heat gun.
- If the temperature difference is greater than 30 degrees, then an alignment adjustment is required.
- Repeat above steps until temperatures of the guide lug faces are similar on both sides.

8.2.2 Setting the tracks

- For adjusting the tractor tracks, first raise one side of the tractor on the jack and keep a rated jack under the jacking points fore and after on the machine.

- Remove the locking bolt and locking plate from both inside and outside of frame and remove the cover plate.

- Loose adjustment bolt on the cold side by 1/2 turn and tighten adjustment bolt on hot side to 200 N-m (150 lb–ft).
- The plate will allow you to access special grease zerk connected to cylinder which increases in length, from where, you can stretch or tighten the track.
- Add or remove the grease gun to add grease to tighten the track. Also, remove the grease to loosen the track.

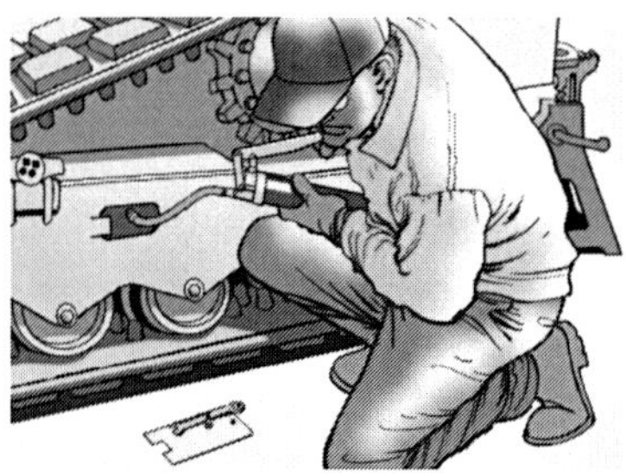

- Now measure the track sag from top to bottom of the track rollers which should be about 1 inch as a default adjustment for optimal usage.

- After this, change or repair the cover plate of lower machine and similar adjust on the opposite side of the tractor track.

8.3 Types of implements & their use

8.3.1 Categories of agriculture implements

In general, the agriculture implements are divided into following categories based on their usage:

- **Irrigation machinery:** It helps to keep crops watered at appropriate levels which include central pivot irrigation systems and pump units.
- **Soil cultivation implements:** It is used to plow the soil and prepare it for cultivation which includes spike, drag, disk harrows, power harrow parts, mouldboard plows and tillers.
- **Planting machines:** It is designed to plant saplings and seeds over broad areas of land, after the soil has been cultivated, which includes precision drills, broadcast seeders, seed drills, air seeders and transplanting equipment.
- **Harvesting implements:** It is used to bring in the crops once they have reached full maturity, which includes trailers, diggers and pickers.
- **Miscellaneous agricultural equipment:** It is used in corollary agricultural activities such as hay-making, shredding and loading.

8.3.2 Tractor implements

Box Blade

- It is a type of implement which is for leveling and smoothing of uneven ground.
- It is pulled behind the tractor, so it is able to leave a pristine, untouched surface.

Mower/Cutter

- It is used for clearing grass and light brush.
- It is pulled behind the tractor and many models connect to the power take-off (PTO) to turn the cutter blades.
- Some implement are powered by their own independent motors.
- Some implement are better suited for handling refined landscaping work while others are better suited for knocking down bigger brush in rougher territory.

Loader

- It is a tool with many potential uses.
- It is a set of arms that attach to the front of the tractor which allow it to lift large loads into the air.
- It comes with a large bucket in the front which can be filled with dirt, cut brush, fertilizer, etc.
- It is extremely helpful on farms where it can help moving large quantities of manure, mulch, topsoil and more.

Post Hole Digger

- It is essentially a large auger attachment which works with tractors.
- It is extremely helpful in case of large amount of fence line to maintain.
- It is helpful for landscapers who regularly install large quantities of new plants.
- With this, digging plenty of tree holes is much easier.

Rotary Tiller

- ❖ It uses rotating tines which looks like a large auger to turn up topsoil.
- ❖ It is helpful for turning a cover crop into the ground or for preparing the soil for planting.
- ❖ Once the ground is used by the implement, it will have several inches of open, soft soil which becomes ready to accept seeds.

Spreader

- ❖ It is designed to fling a product which could be seeds or fertilizer in wide, even layer around a field.
- ❖ It uses rotating head and centrifugal force to shoot the product to a particular distance.

8.4 Preparing the tractor for field operation

It is the duty of the tractor operator to consider the following, before taking the tractor into the fields:

- ❖ Read and understand decals on the tractor.
- ❖ Make sure that the shield for the PTO and other parts are in place.
- ❖ Keep steps and platform free of dirt, grease, and debris.
- ❖ Check if the slow-moving vehicle emblem is in place, clean, and not faded.
- ❖ Test lights to be sure they're working.
- ❖ Check for loose tools and parts.
- ❖ Clean the windows.
- ❖ Read and understand all controls.
- ❖ Check the condition and pressure of all tires.
- ❖ Check oil level; add oil as needed.
- ❖ Check coolant level; add coolant as needed. Do not remove radiator cap when hot.
- ❖ Check the battery's electrolyte level and add battery water as needed. Don't use matches around the battery and don't smoke.
- ❖ Check, clean, coat and tighten battery connections.
- ❖ Locate all grease fittings and clean and lubricate them.
- ❖ Refuel the tractor, making sure that the engine is cool and refrain from smoking.
- ❖ Adjust wheel width.
- ❖ Add or remove weights.
- ❖ Check if equipment has been properly serviced and adjusted.
- ❖ Starting, Operating, Stopping Tractor and Equipment
- ❖ Make sure that the area is clear of people, pets and obstacles.
- ❖ Adjust seat for comfort.

- Start the tractor and check gauges for proper readings.
- Listen for unusual sounds and shut off the engine if you hear any.
- Check brakes.
- Warm engine before applying a heavy load.
- Raise, lower and extend the drawbar.
- Clean and oil PTO shaft and splines.
- Clean, connect and disconnect hydraulic lines.
- Connect and disconnect electrical connections.
- Check hydraulic controls for proper operations.
- Engage PTO slowly; check for proper operation.
- Operate equipment at proper forward speed.
- Operate equipment at proper PTO speed.

8.5 Adjustment of clutch & Brake

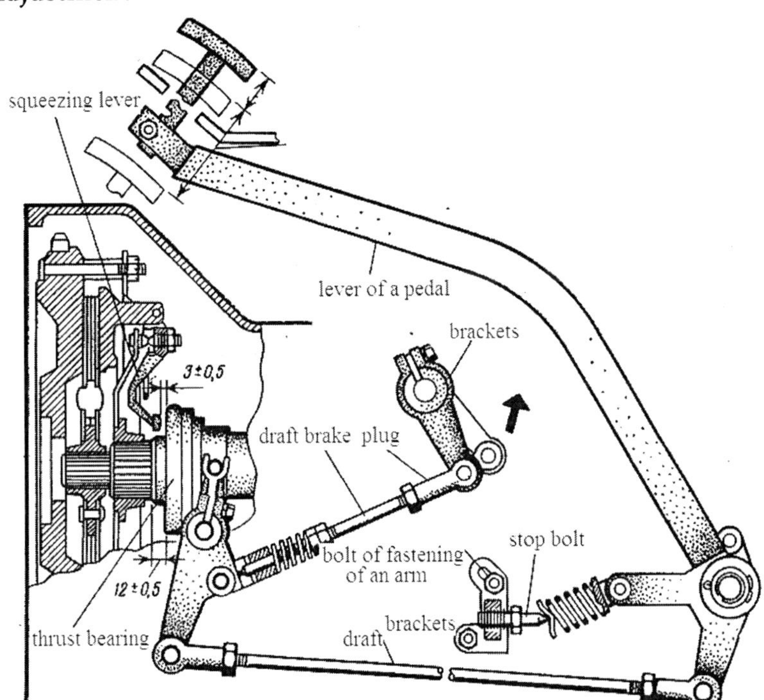

8.5.1 Adjustment of clutch

Tractor mechanic/operator should adjust the clutch, by using the following instructions:

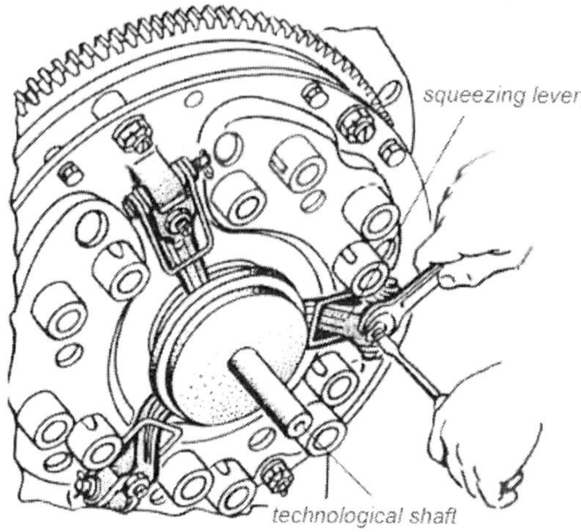

- The free travel of the clutch pedal is controlled by the rotation of the linkage.
- In the released position, the pedal lever should rest against the cockpit polish.
- If lever is not resting against the cockpit polish, then slowly remove the stop bolt from the bracket.
- Also, loose the bracket mounting bolt and turn the bracket clockwise towards the spring.
- If the clutch release mechanism is correctly adjusted and the dimension is 12 ± 0.5 mm, then the gap between the thrust bearing and protrusions of the levers will result as 3 ± 0.5 mm.

8.5.2 Adjustment of brake

Tractor mechanic/operator should adjust the brake, by using the following instructions:

- Disconnect the draft from an arm
- Turn the bracket to the right anticlockwise until it stops
- Rotate the fork, which increases the length of the linkage till the bracket and plug fit freely.
- Rotate the plug till the total length of the rod is reduced by 7 mm
- With this position, connect the plug to the bracket
- Once the adjustment is done, screw the plug with the nut

Unit 2: Tractor Maintenance

Chapter 9

Study and Follow Owner's Manual

1.1. Follow intervals for routine maintenance

Tractor plays an important role in mechanised agriculture. To keep the tractor working for longer time at minimum cost, it requires an adequate and timely maintenance. Tractor operator should do and follow regular maintenance of tractor and its related parts by following the routine and periodic maintenance which are based on the number of actual hours of tractor operations. Regular and periodic maintenance can be easily carried out by maintaining a schedule of operation which will remind tractor operator to regularly follow-up the maintenance activities.

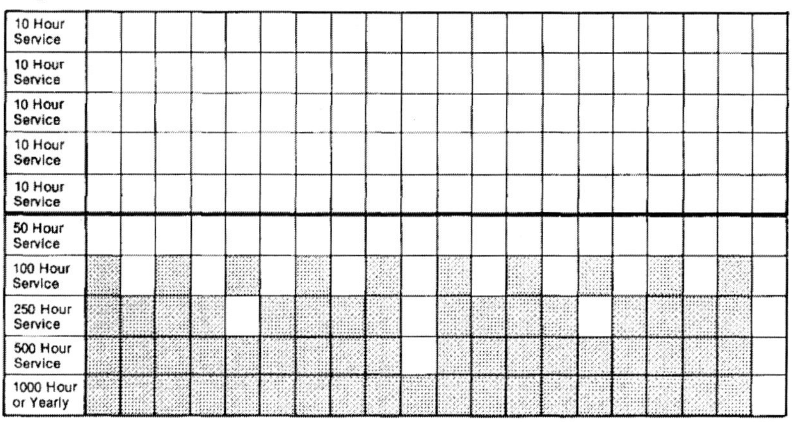

Fig Periodic Maintenance details based on working/operational hours

For this, tractor operator has to read the maintenance manual as provided by the tractor manufacturer which describes the required maintenance procedure and schedule as per the number of tractor running hours which covers maintenance after every:

- 10 operating hours.
- 50 operating hours.
- 100 operating hours.
- 250 operating hours.

- 500 operating hours.
- 1000-1200 operating hours.

Tractor operator has to regularly follow the maintenance of tractor and related parts based on the chart and should regularly update the maintenance details for future follow-ups.

a. Maintenance after 10 operating hours (daily)

There are certain tractor checking and maintenance activities which need to be done after 10 working hours of the tractor which includes:

1. **Checking Air Filter:** In tractor, filter protect against dirt, water or other contaminants that could cause failure of the components. Before starting for work, check the air filter and clean it with a shop vacuum or with compressed air.

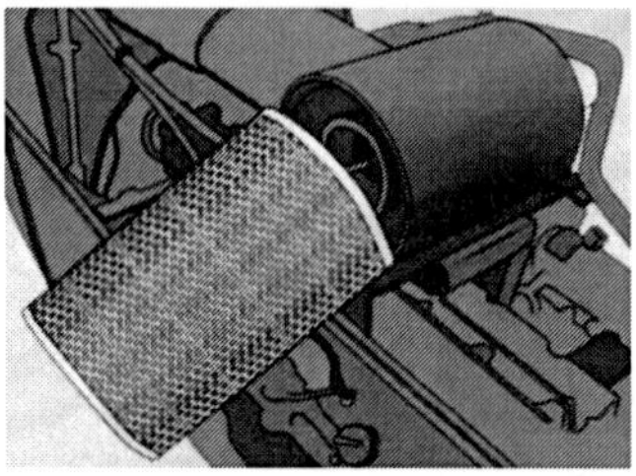

Steps:

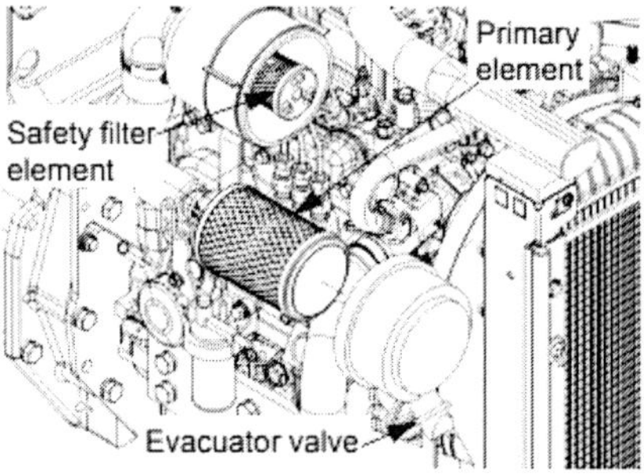

- Remove the cover and pull the primary element straight out, ensuring the safety filter element intact.
- When cleaning the element in working field, clean element by hand to remove the dust
- If the dust is not removed, use the compressed air.
- Check if there is damage inside filter element using a light. In case of crack/hole in filter element or gasket, change it.
- Insert the filter element by pushing it deeply into the filter housing.
- Remove the dust of evacuator valve and clean the inside of cover.
- Assemble the cover with the evacuator valve which is downward

2. Checking Oil Level:

Engine Oil level

Tractor operator should check the tractor engine oil by inserting an oil dipstick in the engine and then pulling it up, to have an idea about the oil level as indicated on the dipstick. In the dip stick, high and low marks are present, which shows an exact oil level.

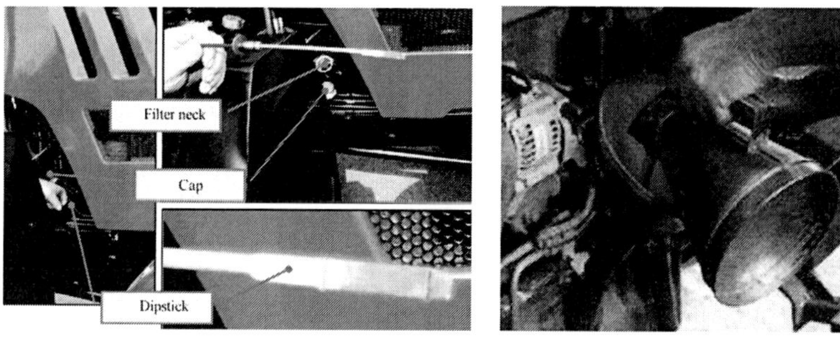

Fig Checking oil level and filling oil

Transmission Housing Oil Level

- ❖ Visually check the oil level on oil-level indicator (1) which is located on the left side of transmission system.
- ❖ The oil level shall be within 10 mm from the mark "П" ("F"–full).
- ❖ If necessary, remove the plug of the oil filler neck and add oil within ± 5 mm from the mark "П".

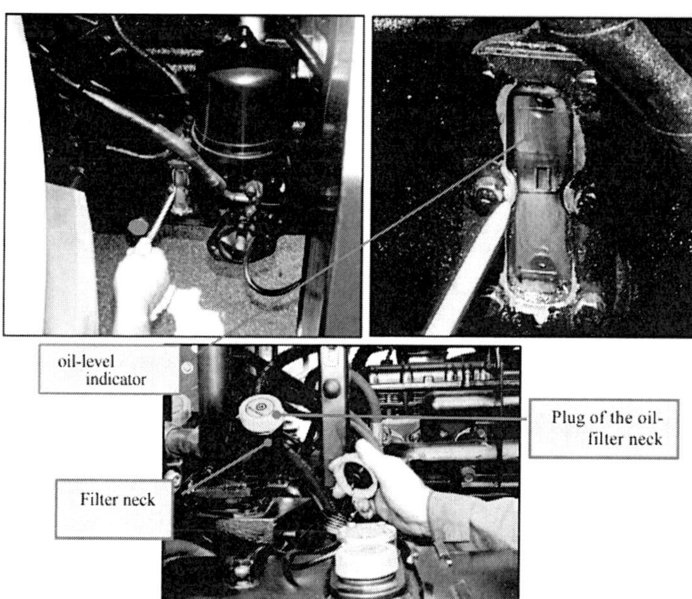

3. Checking Cooling System Level

When the tractor becomes heated up, you need to check for the radiator. You can cool the tractor radiator by water. Do not allow water level to go below from the top of the radiator.

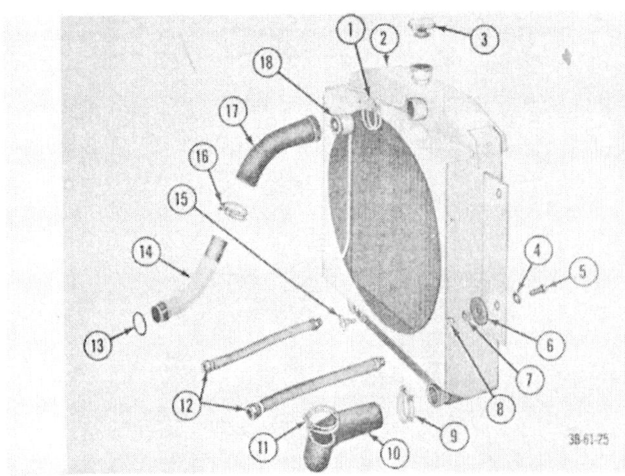

1 Clamp, hose
2 Radiator
3 Cap, pressure
4 Washer, plain
5 Bolt
6 Washer
7 Washer, plain
8 Locknut
9 Clamp, hose
10 Hose, radiator outlet
11 Clamp, hose
12 Hoses, transmission filter outlet and cooler
13 Packing
14 Tube, cylinder head
15 Cock, drain
16 Clamp, hose
17 Hose, radiator inlet
18 Thermostat

Once the radiator gets cooled, check the coolant level by removing the radiator cap. Coolant should be levelled with upper end-face of filler neck and if not, add the coolant to the level.

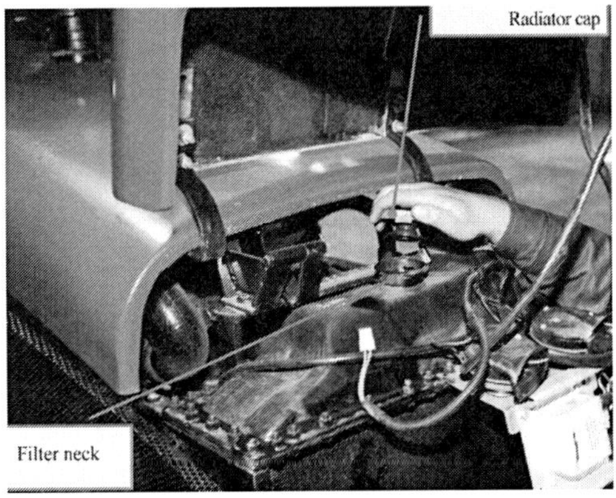

4. Lubricating Grease:

- Good lubrication of tractor parts ensures reduced maintenance, optimised operating costs and extended lifetime for equipments.
- Tractor parts should be greased properly for smooth running and increases working efficiency as dust will make the parts harder.

GREASING POINTS ON TRACTOR

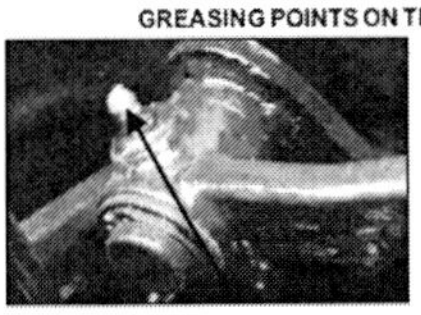
Center Pin (01pt.)

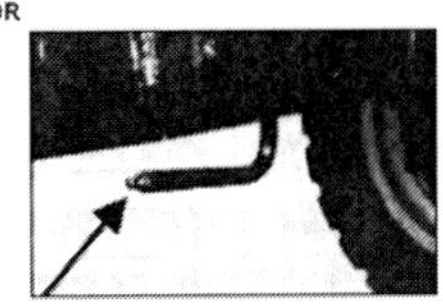
Center Pin

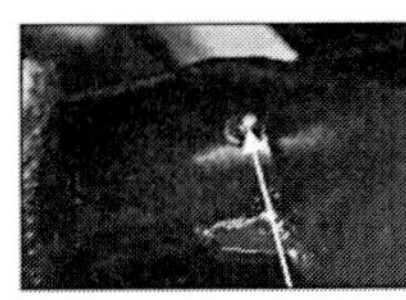
King Pin (Left / Right)

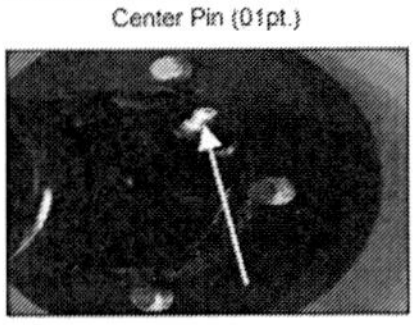
Front Wheel Hub (Left / Right)

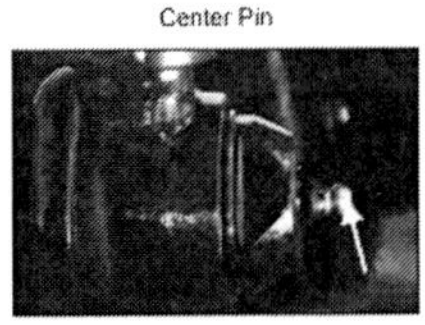
Tie Rod (02pt.) (Left / Right)

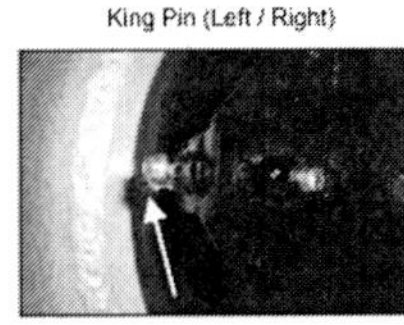
Trumpet plate cover (left(01pt.) Right (02pt.)

Steering Arm (Left)

Steering Arm (Right)

Lift rod (Left lift rod (01pt) & Right Lift Rod (01pt.)

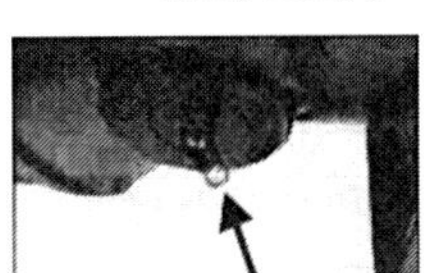
Clutch actuator shaft (left (01pt.) Right (01Pt.)

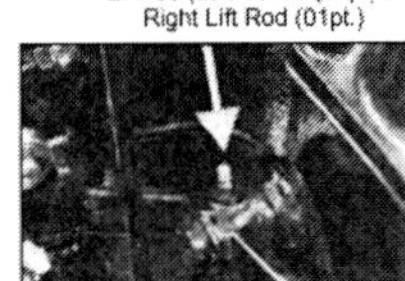
Differential pedal (01pt.)

5. Checking Fuel Sediment Bowl

In tractor, fuel sediment bowl assembly is located in between source of fuel and fuel pump.

The fuel sediment bowl assembly is designed particularly for separating water from fuel system. For checking the presence of water in the bowl, keep the plastic ring in the bowl and allow it to float. As diesel is lighter than water, so when there is water in the fuel, the plastic ring will float.

6. Checking Loose or Damaged Parts

Tractor operator, apartment from checking should also examine the tractor to look for any loose or defectives parts that could be frayed or worn fan belt. Tractor operator should replace, tighten or make necessary repairs before heading out in the field next day.

7. Checking for Leakages:

Before working on next day with the tractor, tractor operator should quickly examine for any leakage related to fuel, oil, coolant, air pressure etc.

b. Maintenance after 50 hours of operation: -

There are certain tractor checking and maintenance activities which tractor operator needs to be done after 50 working hours of tractor which includes:

1. Repetition of 10 hour service schedule

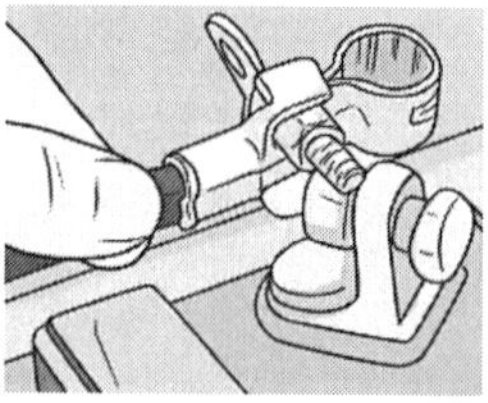

2. Maintenance of Tractor Battery

a. Checking battery terminals:

❖ Check the battery terminals and fittings to make sure there is no corrosion or loose connections.

❖ If any corrosion on the battery holder or in framework is seen, then remove the battery and clean the rack and the holder with the brush.

1. Indicator Light 2. Wing Nut Clamp
3 . Battery Cover

b. Checking battery water:

❖ Battery has cell compartments. Battery water is required for charging, so in case the level of water is below the indicated line, then top up the distilled water to maintain the level.

❖ Do not use acid instead of battery.

c. Battery testing:

❖ Battery testing should be considered as main part of periodic maintenance routine and should be performed whether or not a starting problem has occurred.

❖ Before testing a battery, it is important that the battery is fully charged.

❖ A battery tester is used which shows the life and condition of the battery.

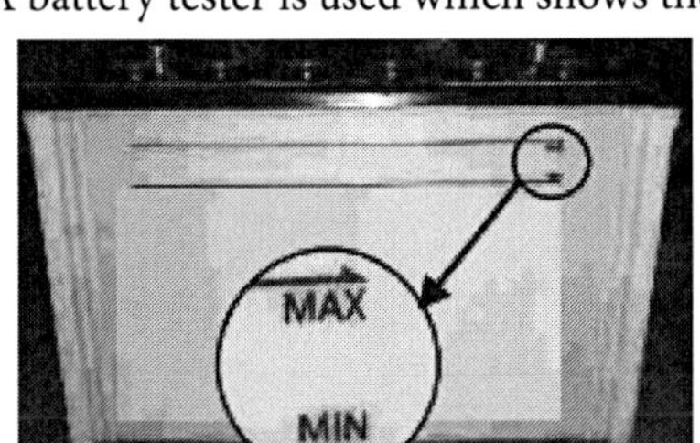

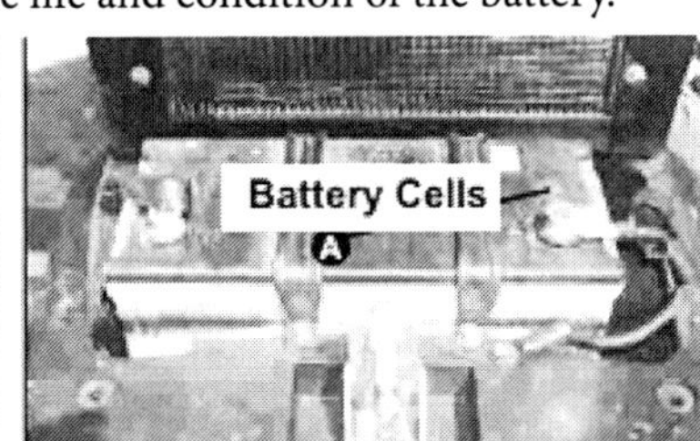

3. Check fan-belt tension:

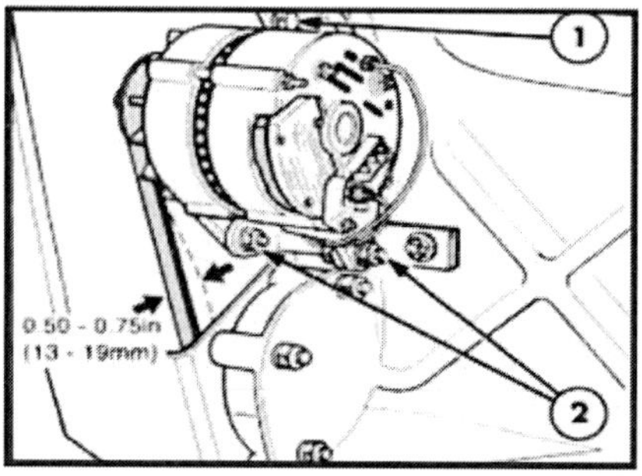

1.Alternator mounting bolts 2 Belt tension

- ❖ The fan/generator belt serves as automatic belt tensioner.
- ❖ While checking belt tension, make sure that belt is in good condition.
- ❖ A slack, worn or oily belt can cause problems with battery charging and cooling system.
- ❖ If the tractor is having a compressor, then also check the tension of compressor's belt.

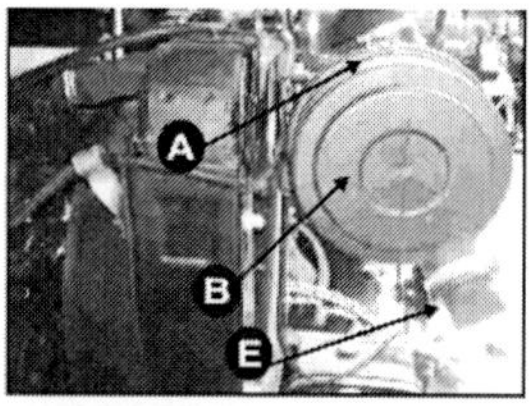

4. Cleaning/servicing of air cleaner:

- ❖ After prescribed working hours of tractor, check the air cleaner. If the air cleaner needs cleaning due to dust, then clean and wash it using kerosene/diesel.

- ❖ If the air cleaner is not clean, it starves the engine of air which causes power loss, fuel wastage and engine wear/tear.

To replace the air cleaner elements:

- ❖ Open the bonnet hood.
- ❖ Release clips (A).
- ❖ Remove cover (B).
- ❖ Remove primary element (C).
- ❖ Clean out any dirt in canister by carefully handling of secondary filter element.
- ❖ Removal of secondary element (D) is similar to removal of primary element.
- ❖ Install secondary element, primary element and cover.
    - A—Clips
    - B—Cover
    - C—Primary Element
    - D—Secondary Element
    - E—Unloader Valve

5. **Checking of fuel supply system**:

- ❖ Checking of fuel system involves checking fuel line for leakage and cleaning of sediment bowl and screen.
- ❖ Sediment bowl allows water to settle out and to catch particles of scale rust and other foreign material.

Moisture causes damage by promoting rust, so it needs to be clean by using following procedure:

- ❖ Close fuel cock on fuel supply line
- ❖ Loose the nut that holds the sediment bowl
- ❖ Remove bowl with twisting motion
- ❖ Remove gasket
- ❖ Remove strainer screen or filter
- ❖ Wash screen or filter element with diesel.
- ❖ Clean sediment bowl by washing with diesel or kerosene
- ❖ Open fuel cock and observe flow
- ❖ Reinstall gasket, strainer and sediment bowl
- ❖ Tighten bowl against gasket
- ❖ Open fuel cock by starting the tractor for checking of leakage

6. Check and adjust the brakes/pedal

Brake is an important part in a tractor, which should be checked regularly. Due to unbalanced play at left and right pedals, one-sided braking system gets affected. Keep the similar play on both pedals.

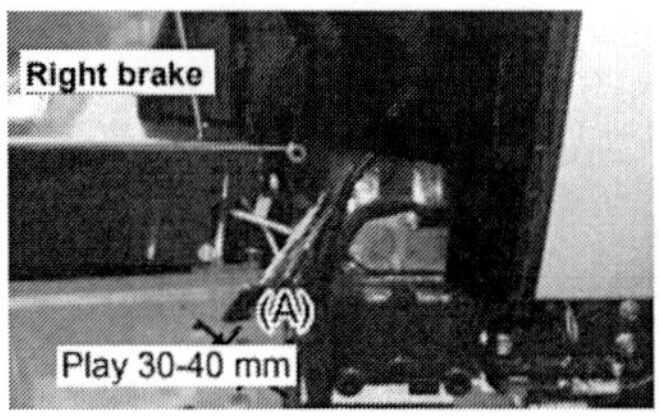

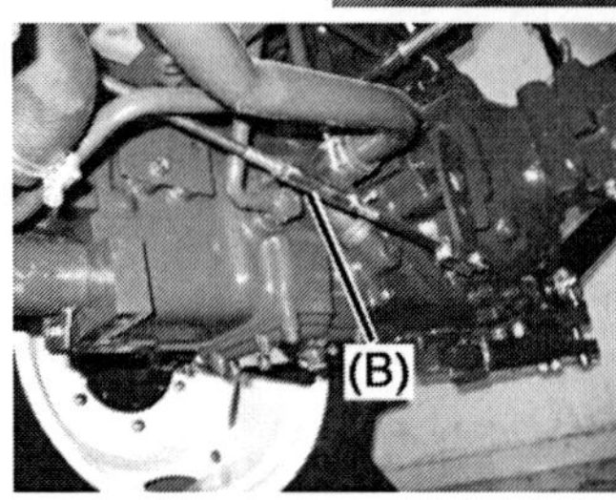

For checking the brake:

- ❖ Depress the brake to see required play of 30-40 mm for left/right pedals to work together.
- ❖ If not, then adjust the turnbuckles behind the pedals for a play of 30 mm.
- ❖ Secure the turnbuckle with lock nuts.
- ❖ Make sure the parking brake lock is put on when the brake pedals are fully stepped in.

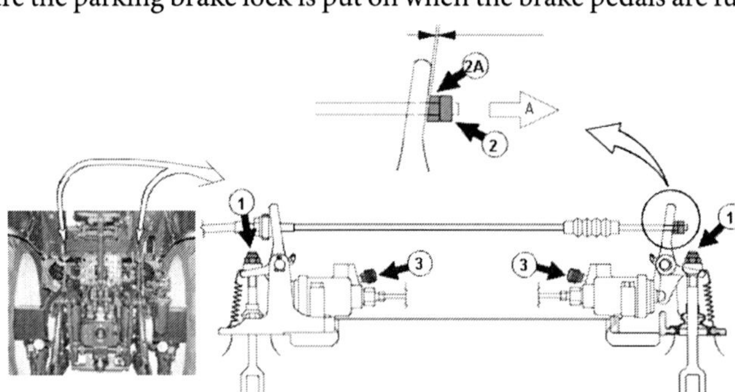

Adjustment of brake

- ❖ These brakes are controlled with a spring return pressure cylinder which is connected to foot brake mechanism by a cable.

- ❖ The parking brake is adjusted in factory and re-adjustment is not necessary till parts of brake mechanism have been changed.
- ❖ While adjusting, parking brake should be off.
- ❖ The tractor engine must be running, shuttling main gear and range levers in neutral position.
- ❖ Fit the blocks in front of the front wheels to prevent the tractor from moving.
- ❖ Pull the cable clearance to the end of the cable.
- ❖ Open the locking nut (2) and adjust through the adjusting nut (2A) the tolerance is 1-2 mm.
- ❖ Tighten the locking nut.

Checking Brake Fluid

- ❖ Visually see the brake liquid levels in reservoir (3) of master cylinder of clutch which is on left side as seen from front of tractor above the hydraulic system oil tank and in the reservoirs (1, 2) of master brakes on right in tractor motion, over HPS oil tank
- ❖ The level should be within marks "min" and "max" on the boxes.
- ❖ If necessary, add the brake fluid to "max" marks having unscrewed preliminarily the caps (3).

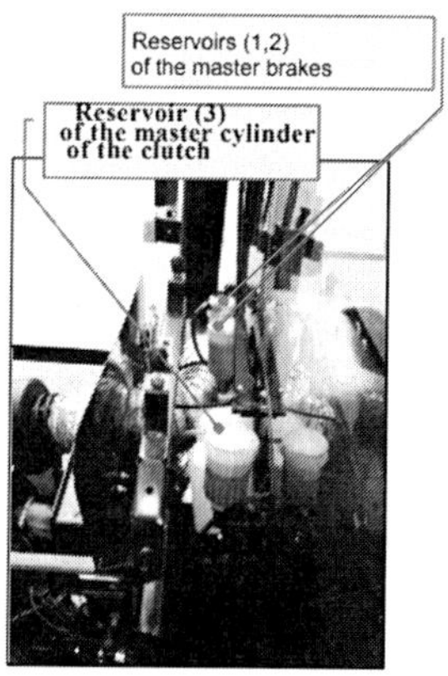

7. Check and adjust the engine clutch:

Tractor operator should check and adjust the tractor engine clutch as:

- ❖ Clutch in tractor operates for much longer period and is commonly in use by the driver while driving.
- ❖ The clutch operates by foot that rest on clutch pedal.
- ❖ The clutch pedal should be adjusted to give a free play of 10-15 mm.
- ❖ The setting is obtained by adjusting the screw until the required measurement is obtained.

- ❖ If the clutch is not free in moving, it will lead to clutch slip, clutch overheating and rapid wear of lining materials on clutch plate.
- ❖ Once the clutch start slipping it will wear out very rapidly, so it is required to adjust it to avoid wear/tear.

8. Checking Transmission oil

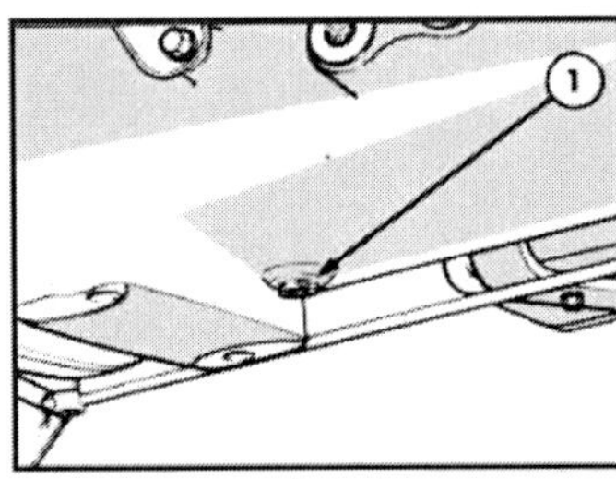

1.Drain plug

- ❖ With the oil is hot, remove the drain plug from the bottom of transmission housing and allow the oil to drain.
- ❖ Unscrew and remove the filler plug to allow faster drainage.
- ❖ Replace the drain plug.
- ❖ Refill the transmission with clean oil.
- ❖ Pull out the dipstick and wipe it with a clean cloth.

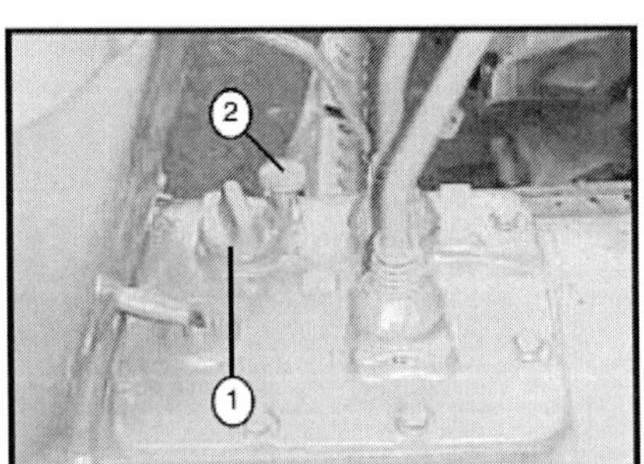

1. Breather cumfiller 2. Dipstick

- ❖ Replace the dipstick, pushing it fully in and check that the oil reaches the upper mark on the dipstick when it is removed from the transmission aperture once more.

- ❖ After refilling the transmission, start the engine to allow the oil to circulate throughout the transmission.
- ❖ After a minute, stop the engine and recheck the oil level by means of the dipstick.
- ❖ If required, top-up the oil to the upper mark on the dipstick.

9. Checking Front axle oil

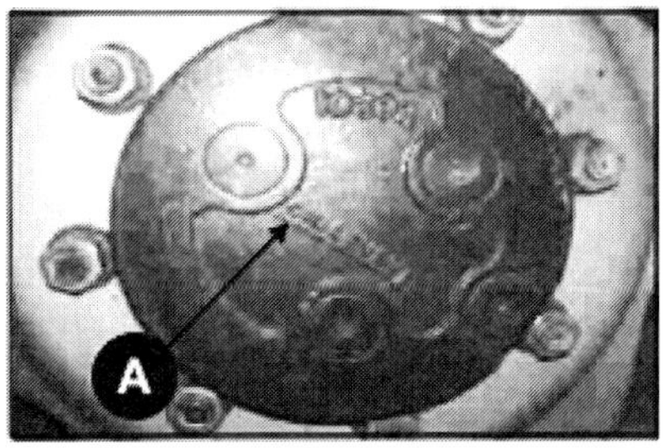

- ❖ Unscrew the oil gauge and check if the oil level is between low and high marks.
- ❖ If the oil level is low, add new oil in oil filling inlet.
- ❖ After 5-10 minutes, again check the oil level.

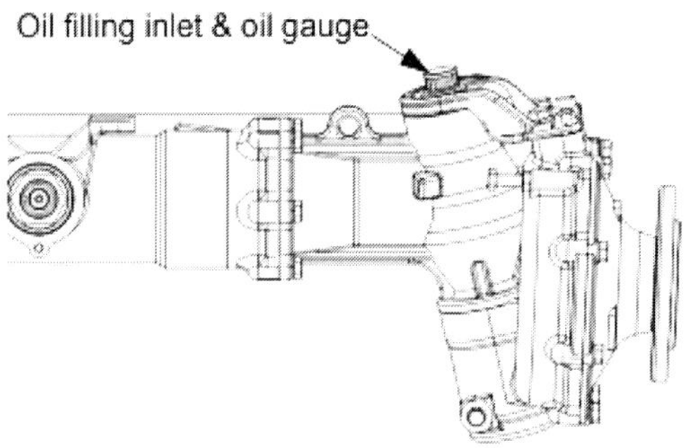

10. Checking of Hydraulic hoses and Leakage

- ❖ Stop the engine of tractor and keep all transmission gears in neutral and further lower down the implement to the ground.
- ❖ Periodically check he hydraulic system for leakage or damaged parts which could be kinked, crushed, flattened, hard blistered, heat cracked, charred, twisted, soft or loose covered hoses and fittings.
- ❖ Before removing hydraulic components, make sure to check that hydraulic pressure is relieved completely.
- ❖ The leakage of pressurized oil could result in fatal physical injury.

11. Checking Oil Level in Oil Reservoir

- ❖ Visually check the oil level against the oil-level indicator located on HPS tank placed on right-hand side on clutch casing.

- ❖ Here, the level indicated should be between «O» and «П» as marked on the gauge.
- ❖ If the level of oil is low, add oil.
- ❖ For adding oil, remove the plug, together with the valve and add oil up to the mark «П».

c. Maintenance after 120/125 operating hours (Fortnightly)

1. Repeat 10 and 50 Hour Maintenance

2. Greasing of Parts

Pivot Points of HPS Cylinders

- ❖ After 100 operating hours, the user needs to perform oiling and greasing activities through manual operations or by using lubricating gun.
- ❖ Grease the pivots through the lubricating plugs (2) in four lubricating points.

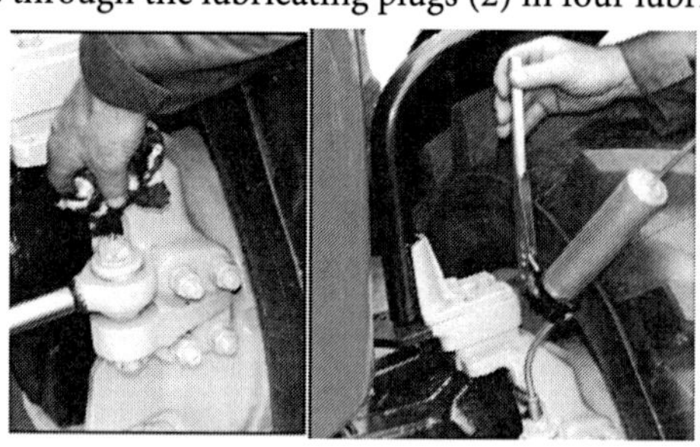

Bearings of Pivot Supports of Hub Reduction Gear and Pivot Centre Bushings

- ❖ Use the gun greasing approach to lubricate (1) 4 to 6 strokes points.
- ❖ Gun-grease the lubricator (2) with required lubricant till fresh grease is squeezed from the gaps.

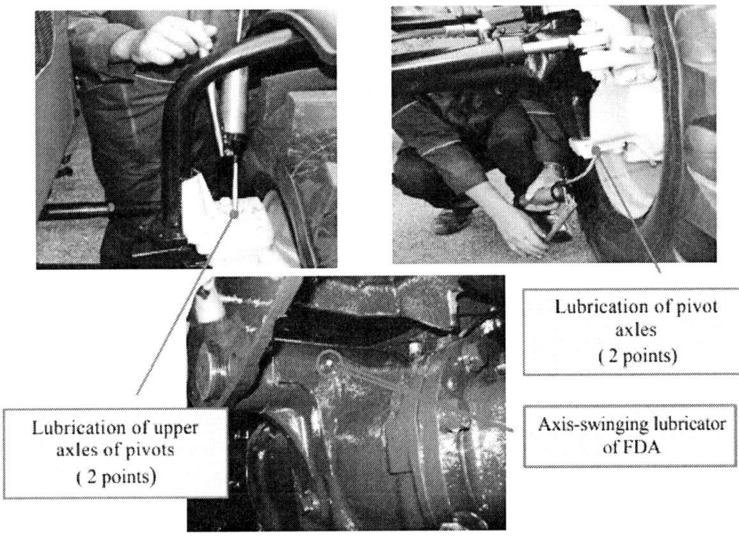

3. Cleaning of fuel tank

- Unscrew the plugs and drain out the sediment and sludge from the fuel tanks and filter.
- Remove the unclear fuel from the tank.
- Screw in the plugs.
- Refill the tank with fresh fuel

4. Checking Tension of Alternator Driving Belt

- Alternator generates current which charges battery for good electrical backup.

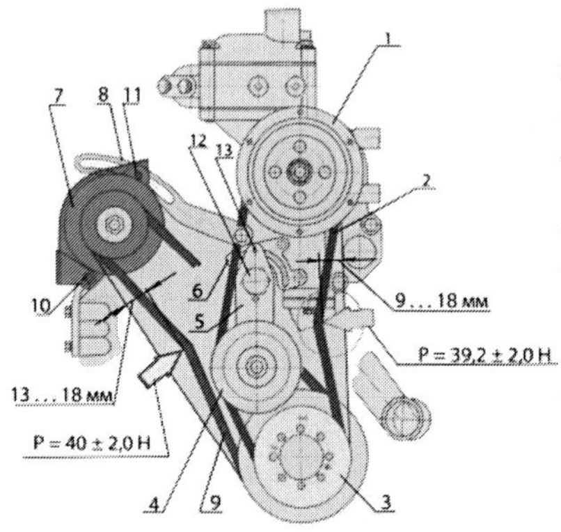

1 – water-pump polley;
2 – water-pump drive belt;
3 – crankshaft pulley;
4 – water-pump belt tentioner idle pulley
5 – lever;
6 – plate;
7 – alternator;
8 – plate;
9 – belt;
10 – bolts;
11 – bolt;
12 – axle;
13 – bolt.

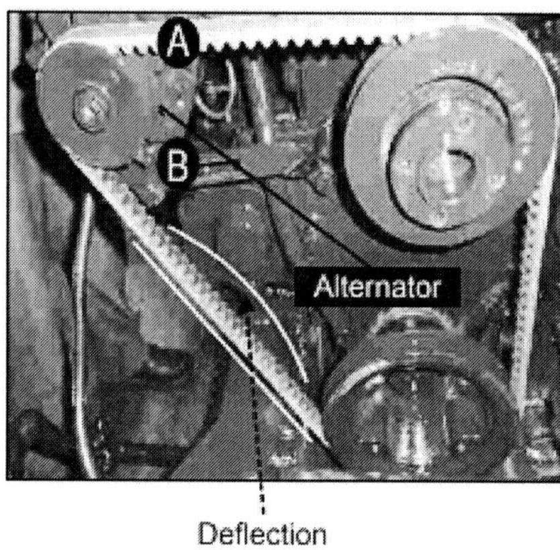

- ❖ Tension in belt is considered correct when sagging at section among crankshaft pulleys and alternator is 29-33 mm when pushed due to heavy force.
- ❖ To adjust belt tension, loose the nuts and turn the alternator body for checking and removing the belt.
- ❖ Keep the alternator outwards to obtain proper belt tension.
- ❖ Tight the plate attachment bolt and nuts of alternator (A and B).

5. Adjusting Clutch Control

- ❖ Set the initial position of pedal (7) by means of the adjusting bolt (3) and fork (5) using dimension "B"
- ❖ Adjust the clearance between the piston lifter (9) and piston of master cylinder (10) using fork (5).
- ❖ When doing this, travel of pedal (7) from initial position to moment of contact of piston lifter (9) using piston as measured at centre of pedal boot corresponds to dimension "B".
- ❖ Tighten the nuts (4,8) and fix the pin (6) with a cotter pin.

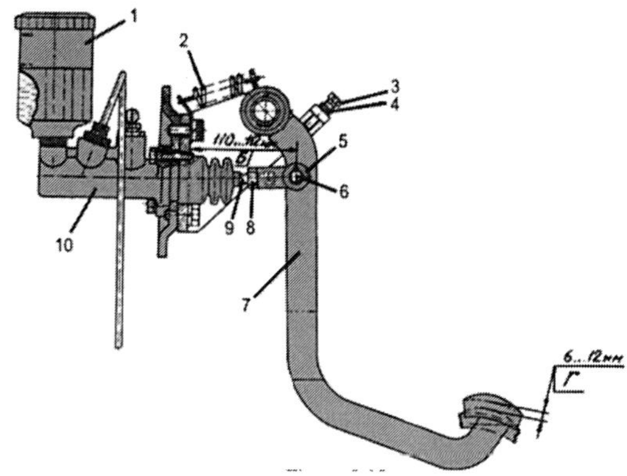

- Move the cover 13 down to make the lock nut 14 available;
- Remove the piston lifter 12;
- Unlock the piston lifter 12 of the maim cylinder 19 in fork 15, with F dimension;
- Tighten the nut 14 and refit the boot 13;
- Adjust the free travel of pedal 20, as measured at centre of boot to dimension F (6…12MM) by screwing/unscrewing bolt 17 with clearance 1,5…2.0 mm among piston lifter 12 and main cylinder 19;
- Tighten the locknut 18;

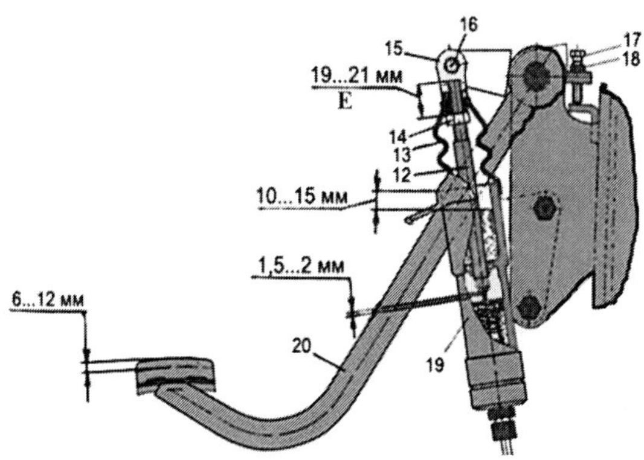

- Remove master cylinder (30) from bracket (33) with pin (34) removed;
- Set piston lifter (31) in cylinder (30) to rightmost position till it rests against the cover;
- Move the cylinder (30) until it contacts piston lifter (32) of hydraulic booster (39) without effort.
- Align the holes of support (26) with bracket (33) by turning the support (26) in or out;
- Screw the support in by 1/2 revolutions and refit the pin (34);

- Tighten the locknut (27) and fix the pin (34) with cotter pin

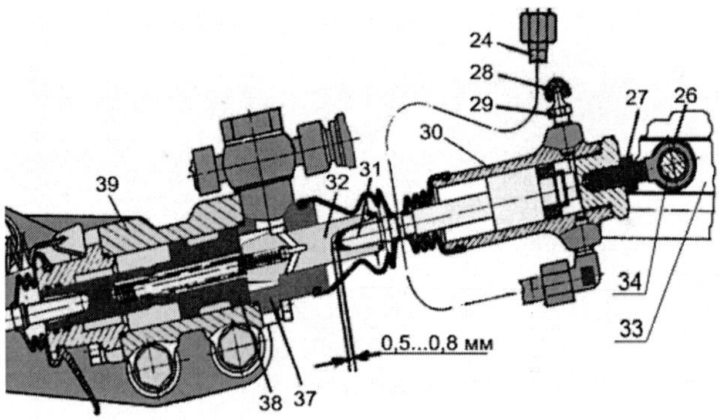

6. Clean the filter

- For checking the filter, lift the panel of cab.
- Turn out two screws with plastic heads under projecting edge of cab.
- Remove the panel and the filter.
- Shake dust out of filter by slight tapping, so that it will not damage the paper filtering element.
- Clean the filter by compressed air under pressure of not more than 0.1 MPa.
- The hose nozzle to be held at distance of not less than 300 mm from the filter to save damage of paper filtering element.
- It is necessary to direct the air flow through the filter in direction opposite to normal air flow as shown by arrows.

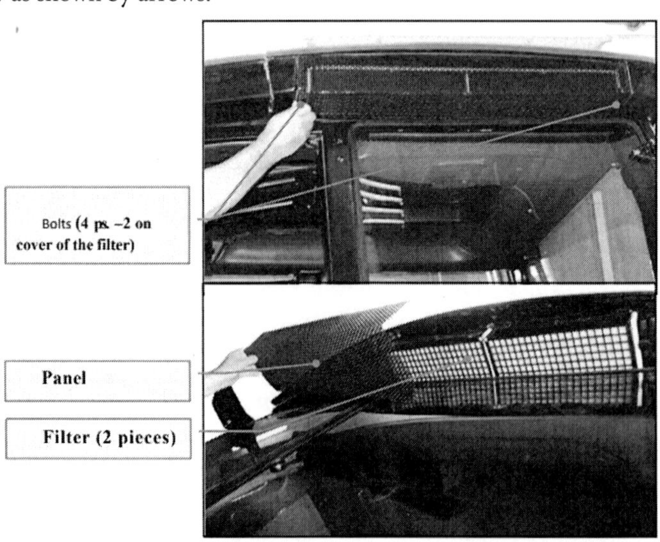

d. Maintenance after 250 operating hours

1. Perform 50 and 100 hour maintenance operations

2. Replacing Engine oil and Filter

a. Draining of Engine oil

- ❖ Run the engine for few minutes to warm the oil.
- ❖ Park the tractor on a level surface.
- ❖ Remove LH and RH drain plugs of oil pan and drain the oil completely

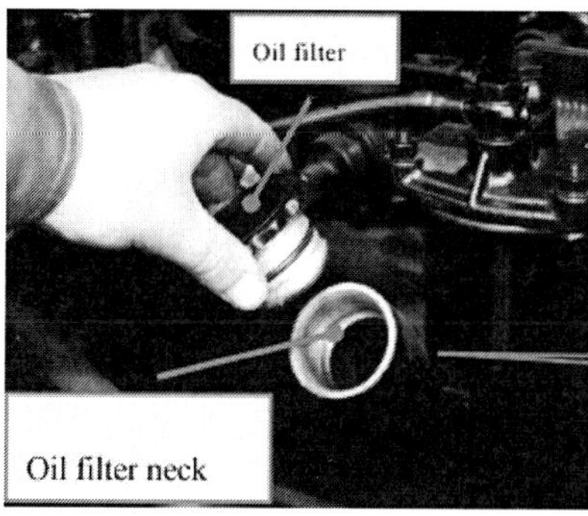

b. Replacing Engine oil filter

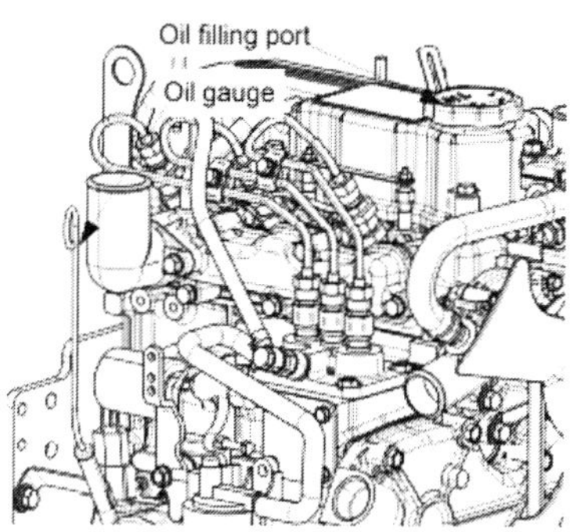

- ❖ Clean carefully around the filter.
- ❖ Coat clean engine oil on the packing of new filter and check for the packing in the groove.
- ❖ Turn the oil filter counter-clockwise to remove with filter wrench.
- ❖ Turn the new filter clockwise to assemble until the packing makes contact with the mounting surface.
- ❖ Tighten ¾ to 1 turn more after packing contact.
- ❖ If the metal is attached to the element of oil filter to be disassembled, contact the local dealer.

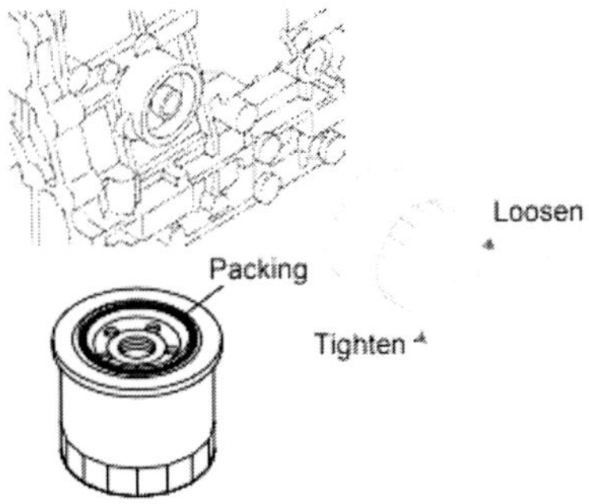

c. Fill Engine oil

- Tight all the drain plugs.
- Add engine oil as its capacity and check the oil level is between MIN and MAX mark on gauge.
- Check for any leakage of engine while running the engine for certain time at idle rpm.
- Stop the engine and after 5-10 minutes, check the oil level again in between MIN and MAX mark.
- Install the oil gauge.

3. Replacing Hydraulic oil filter

- Place the tractor on level surface and apply parking brake and lower implements and stop the engine.

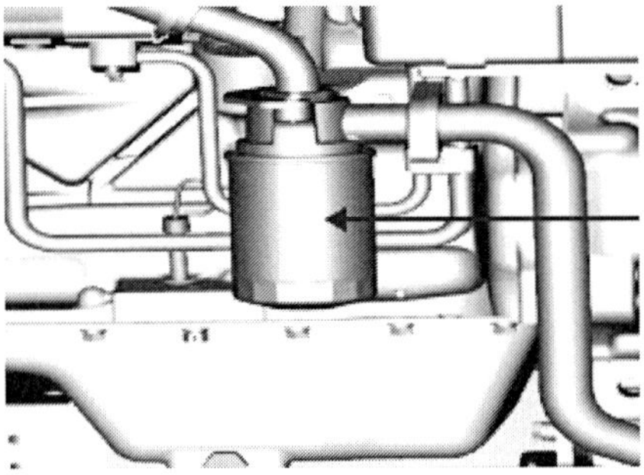

- Clean the filter area and set a clean container under the filter.
- Coat clean hydraulic oil on packing of new filter and check the packing is placed well in the groove.
- Turn the oil filter to remove with filter wrench.
- Turn the new filter to assemble till the packing makes contact with the mounting surface.
- Tighten ¾ to 1 turn more after packing contact.
- Run the engine at idle and check any leakage.

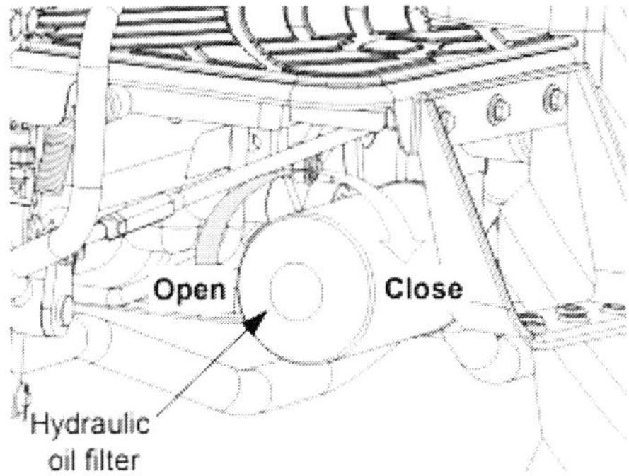

4. Adjust the tension engine belt

When tension of the fan belt exceeds the standard value, then loose the bolt and adjust the tension.

For adjusting belt tension:

- ❖ Loosen two bolts of alternator and hinge bolt of adjusting plate.
- ❖ Insert the bar between the crank case and pull the bar to apply tension to the belt.
- ❖ With proper tension, tighten the bolts attached to the adjusting plate of alternator.
- ❖ If the belt tension becomes normal, tighten the other bolts and check the tightening torque.

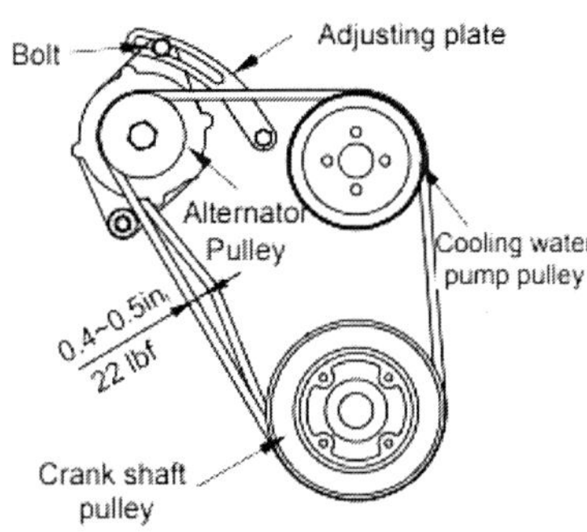

5. Cleaning Rotors of Centrifugal Oil Filters of Engine/Gear-Box

- ❖ For cleaning the rotor, screw out the nut (22) and remove the cap (21).
- ❖ Remove the sleeve (10) using a wrench and screwdriver.
- ❖ Remove the cover (15), impeller (9) and screen (14).
- ❖ Wash the screen (14) in diesel fuel.
- ❖ Remove the layer of sediment from inner walls of the sleeve (10).
- ❖ Grease the O-ring with the oil.

- During assembly align the marks on the sleeve and rotor housing.
- Tight the nut (22)

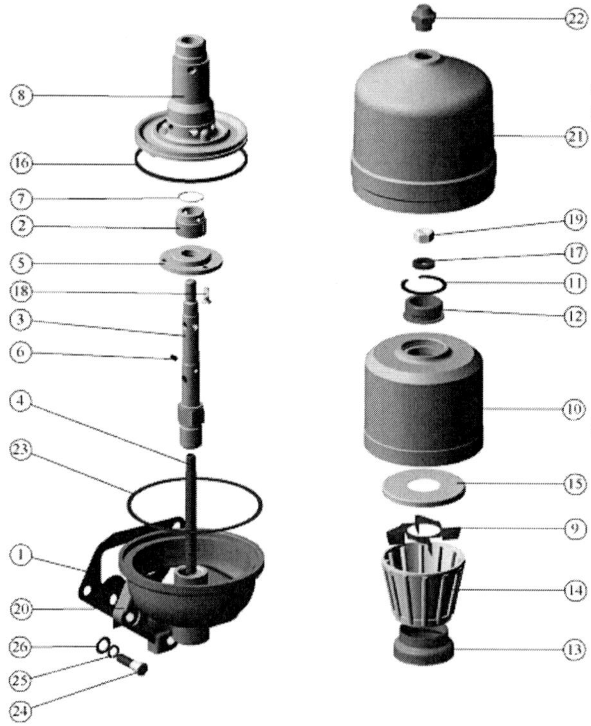

6. Checking Oil Level in Front PTO Reduction Gear-Box

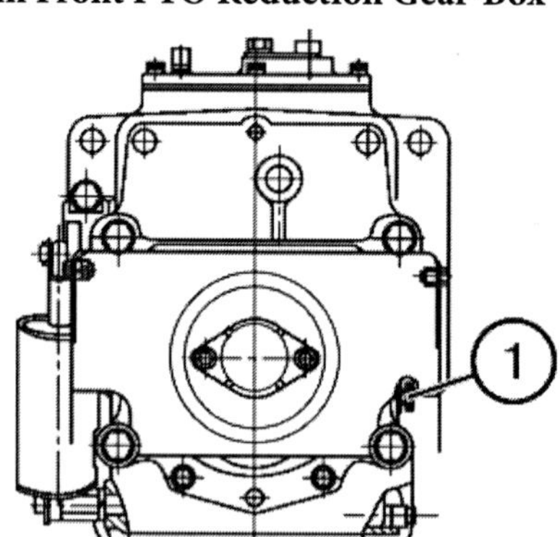

- For checking the oil, unscrew the plug (1) of check-and-filler hole located on right side of reduction gear-box.
- The oil level needs to be till threaded hole for the plug

7. Checking Toe-In of Front Wheels

- ❖ Toe-in is basically the differential distance of back and front measurements from the tyres. It should be around 0 to 8 mm
- ❖ If the toe-in distance exceeds 0 to 8 mm, then adjust it by:
- ❖ Loosening locknuts (1) and (3) of steering link tube (2).
- ❖ Set the required amount of toe-in by rotating the tube in either direction.
- ❖ Tight the locknuts (1) and (3).

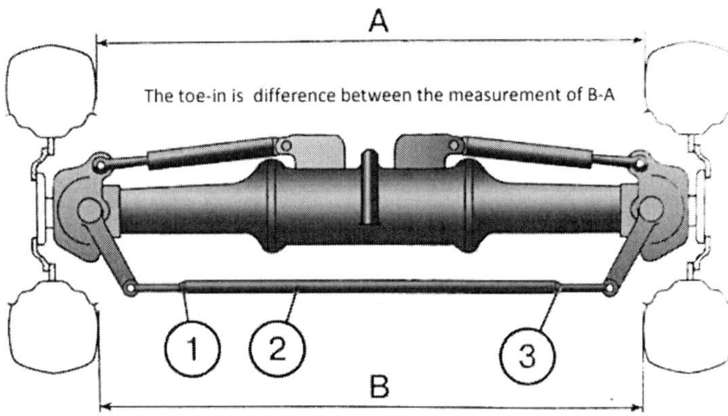

8. Checking Fasteners

- ❖ Due to movement on uneven area, fasteners get loosed.
- ❖ Check the fasteners of the turbocharger (1), outlet ducts (3), exhaust pipe (2) bracket for tightness.
- ❖ Tight the fasteners using suitable wrench.

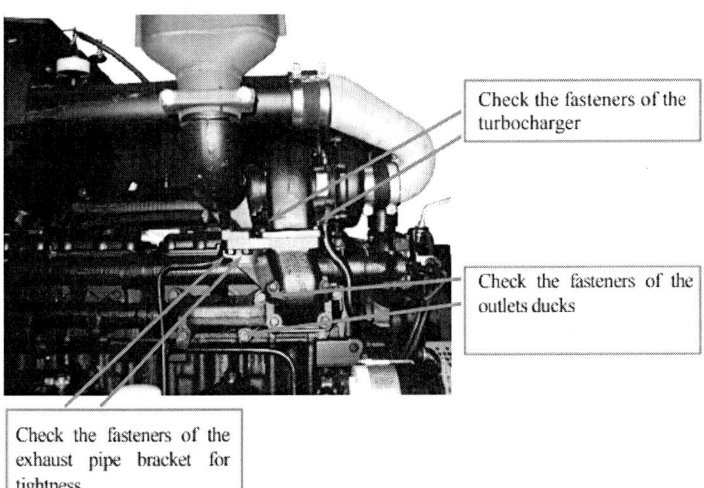

e. Maintenance after 500 operating hours

1. Perform 300 hour maintenance

2. Check/Replace Front axle oil

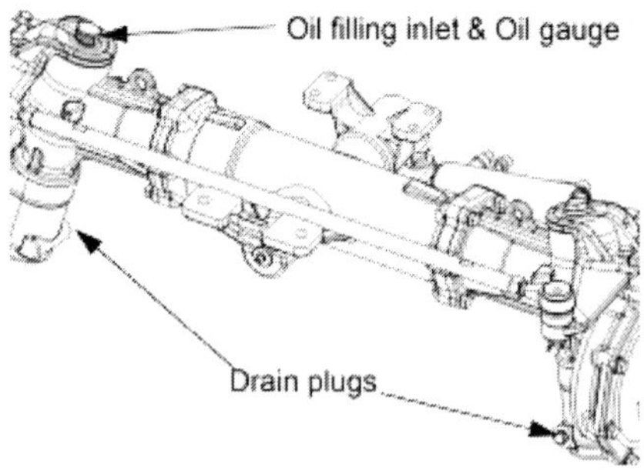

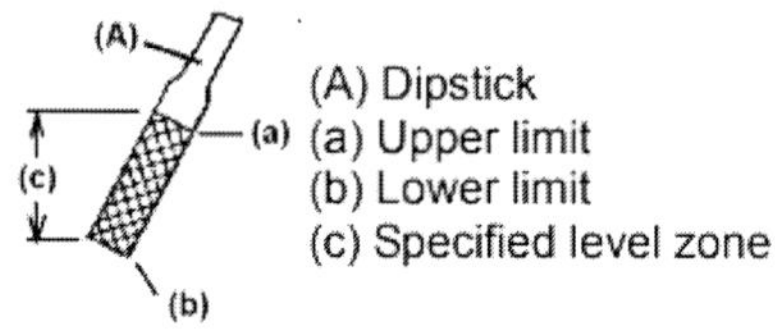

For checking the axle oil:

- ❖ Remove the dipstick plug on the top of the front axle.
- ❖ Clean the gauge rod and reinsert it without screwing.
- ❖ Redraw it and see if the oil level is within the upper and lower marks.
- ❖ If insufficient, add the new oil through the oil supply port to the normal level.
- ❖ Now check for oil leak.

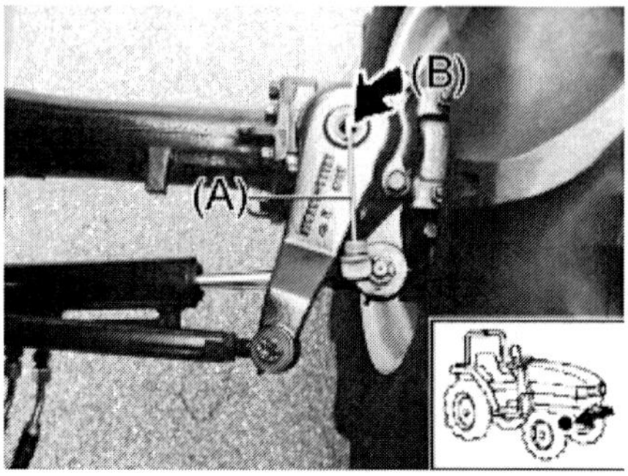

For changing the axle oil:

- ❖ Remove oil through center drain plug on the lower part of front axle.
- ❖ Drain the plugs on the bottom of right and left gear cases.
- ❖ Add oil through the right port.

3. Changing Transmission oil

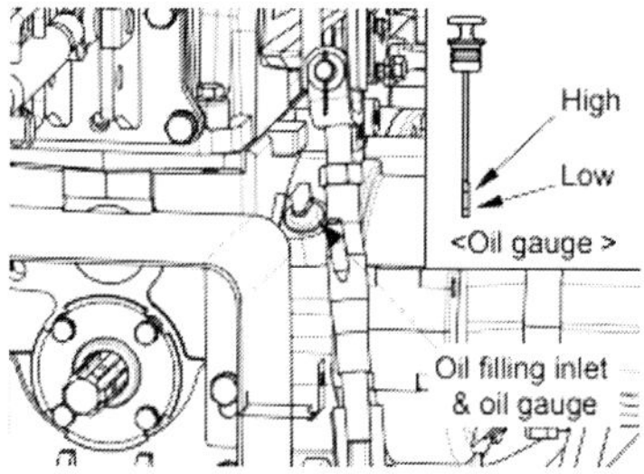

For changing transmission oil:

- ❖ Stop tractor on a level surface and apply brake.
- ❖ Run engine for certain time to warm the oil and lower implements and then stop.
- ❖ Set container under the drain plugs and remove drain plugs under transmission and completely drain the oil.
- ❖ Clean metal powder stuck to the drain plugs and tight the drain plugs with new cooper seals.
- ❖ Add new oil till oil level is achieved between "MIN" and "MAX" of oil gauge.

4. Change Fuel filter cartridge

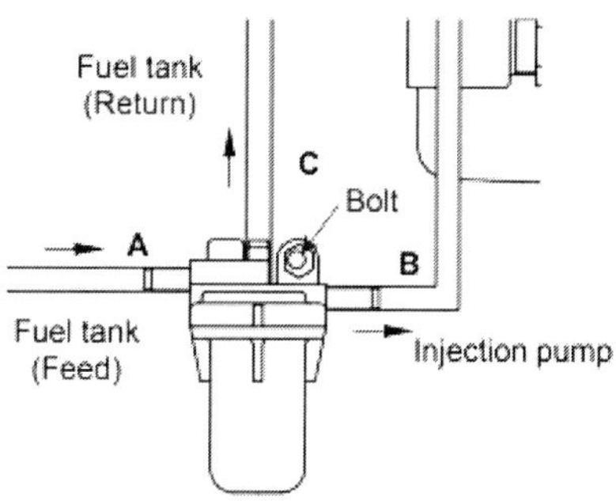

- ❖ For this, stop the engine and apply brakes.
- ❖ Adjust the container under the fuel filter.
- ❖ Prepare the plug to block the fuel hoses.
- ❖ Disconnect hose-A first, and plug the hose.
- ❖ Now disconnect hose-B and C using similar method.
- ❖ Remove the bolt from the filter.
- ❖ Replace the fuel filter with a new one.
- ❖ Connect the hose B, C and A step by step and tighten the bolt.
- ❖ Disconnect hose "C" for a seconds to bleed the air, and connect it again.
- ❖ After this, check the leakage around fuel filter.

5. Checking Bearing Clearances of Front Wheels

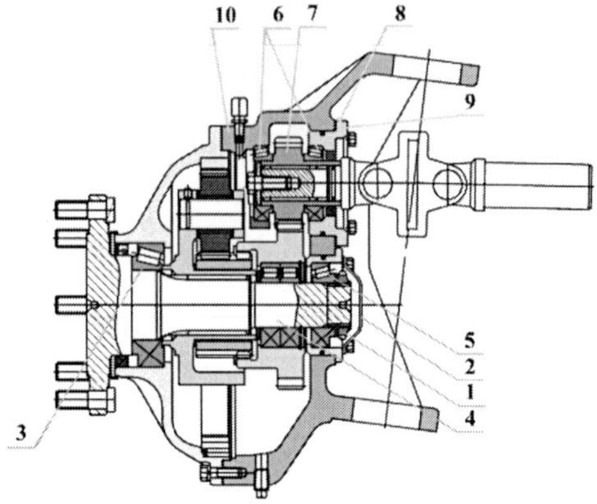

Tractor operator should:

- ❖ check the taper roller bearings (3) of the flange (2).
- ❖ adjust them using the nut to a no-gap condition.
- ❖ tight the nut so that the gap is taken up completely.
- ❖ prick-punch the nut in two slots in the flange (2).

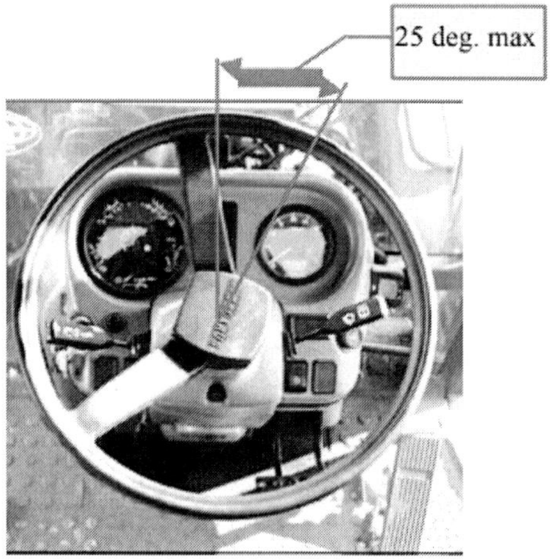

6. Checking Steering Controls

- ❖ Steering is responsible for turning the tractor, so it should be thoroughly checked.
- ❖ It is noted that play in steering wheel should not exceeds 25°.
- ❖ If the play is more, in such case, the steering is loose and its trapezoid pivots carrying nuts, needs to be tightened.
- ❖ By doing this, the play gets eliminated from steering column and steering gear.

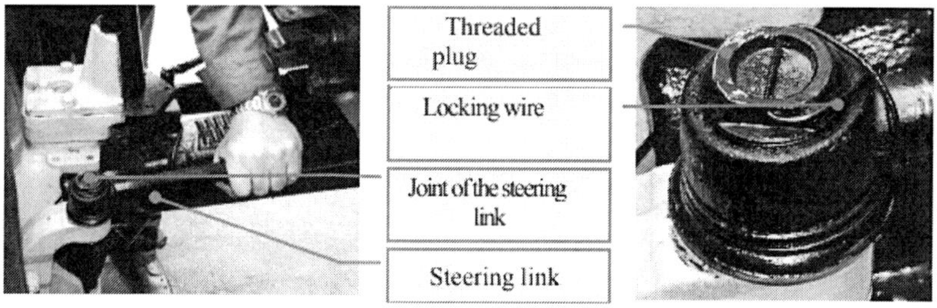

It is noted that when play in the Steering Link Joints cannot be removed by tightening of threaded connections, in such case, disassemble the joint and replace the damage parts.

7. Checking fuel injection timing

Injection timing is timing when fuel is injected inside the cylinder, which gets changed when combustion takes place.

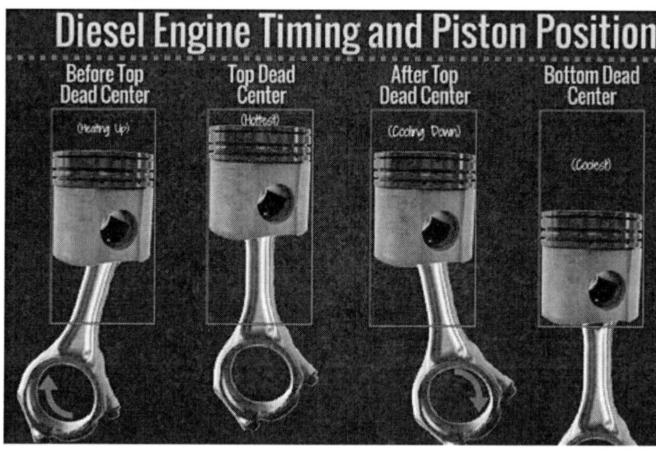

Time at when fuel is injected can be changed for injecting at different points in certain time. The timing is balanced to get as much power as possible. Ways to adjust injection timing includes:

- ❖ programming the ECM
- ❖ adjusting fuel injection pump
- ❖ replacing camshaft
- ❖ replacing cam followers or gaskets

f. Maintenance after 1000/1200 operating hours

1. Follow the bi-monthly maintenance schedule.

2. Change Oil in Transmission Housing

- ❖ Always warm up the oil before changing, as it will not stick.
- ❖ Remove filler neck plug (2) located on right side of clutch housing.
- ❖ Take out drain plugs (1) of transmission and plugs (3) of axle-shaft housings.
- ❖ Take the oil out from the transmission into special vessel for collection and storage of waste oils.
- ❖ After this, fit the drain plugs back in place and fill in fresh oil up to the mark "П" ("Full") mentioned against the oil-level indicator.
- ❖ Also fit the plug (2).
- ❖ Now run the tractor for 5-10 minutes and check the oil level.
- ❖ If required, add the oil level.

3. Changing elements in Fuel Fine Filter

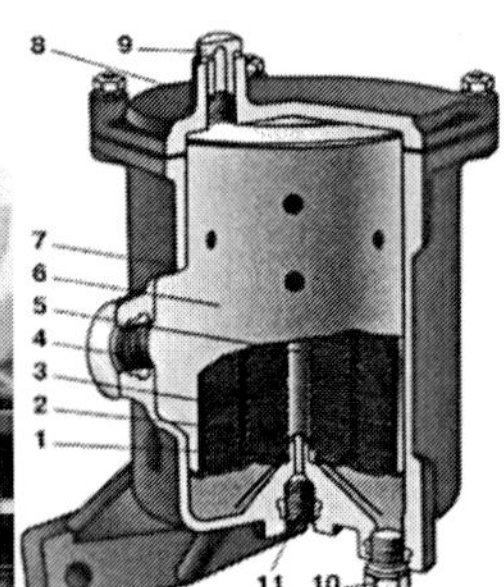

- ❖ For this, unscrew the plug (10) and drain the sludge.
- ❖ Loose the four nuts and remove the cap (8).
- ❖ Take the filter out from the housing and remove the filter element (6).
- ❖ Now clean the housing and cap it with clean diesel fuel.
- ❖ Set the new filter (6).

- Check the cap gasket and change it if required
- Set the cap 8 and tighten four nuts.
- Tighten the plug (10).
- Finally, remove the air from the fuel system.

4. Changing Oil in Gear-Box and Final Drive

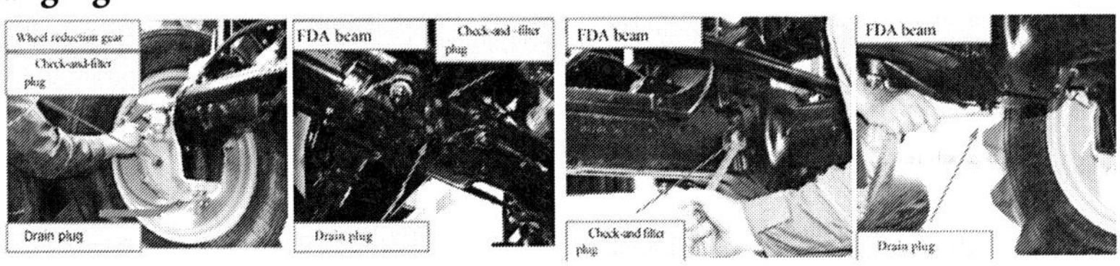

After 1000 kilometers of running of engine, oil in drive and gear box will stop working, so it is advisable to check and replace the oil. For this, follow the steps shown:

- Always warm up the oil before replacing.
- Place the tractor on leveled surface before doing any repair or replacement work.
- Apply parking brake and wedge the wheels on both sides.
- Remove the check-and-filler and drain plugs.
- Remove the oil in a special vessel for collection and storage of waste oils.
- After this, fit the drain plugs.
- Fill the housings with fresh transmission oil to lower edge of check-and-filler necks.
- Finally, fit and tight the plugs.

5. Greasing of Bushes of Turn Shaft of Rear Hitch Linkage/Hauling Mechanism

Greasing after sometimes is really necessary in certain parts of the tractor. Bushes are important part of shaft and linkage mechanism. For doing greasing, follow the steps shown:

- Clean two lubricators placed on the bosses of rear axle cover from dust.
- Clean the lubricator of towbar from dirt.
- Apply grease through gun-grease for more faster and efficient output.

6. Checking the Fuel Injection Equipment

Basically, injection pressure should around 22-25 MPa whose atomization results in form of mist, without continuous jets and/or drips. For checking injectors for pressure, check the fuel injector by removing it from the engine by:

- ❖ screwing out the coupling nuts (6) of unions of fuel pump and injectors;
- ❖ removing high-pressure pipes (4);
- ❖ removing bolts (1) of unions from each injector together with their sealing washers
- ❖ removing draining pipeline (3);
- ❖ screwing out bolts (2) which fix the injectors and remove the injectors (5);
- ❖ now send injectors to specialized workshop or your dealer for checking.

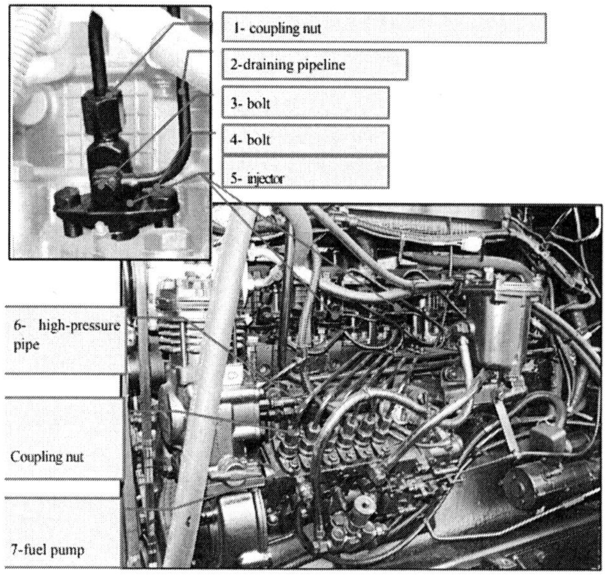

1.2 Lubricate different parts of the tractor

Tractors have many moving parts and it is necessary that these moving parts should be greased regularly to keep them in working conditions. Since many of the parts on a tractor do not feed into a grease reservoir, so they must be greased manually. In order to properly grease the tractor, it is the work of tractor operator to know what parts actually need greasing. Some of the typical parts/areas in tractor which needs lubrication/greasing includes:

- ❖ Ball Joints
- ❖ Bearings
- ❖ Clutch Linkages
- ❖ Master Brake Linkage Shaft and King Pins

- ❖ Steering Cylinder Ball Joints (left and right)
- ❖ Steering Components
- ❖ Three-Point Hitch Pivot Points
- ❖ Tie Rod Ends (left and right)
- ❖ Front Driving Axial

1.2.1 Lubrication of Bearings

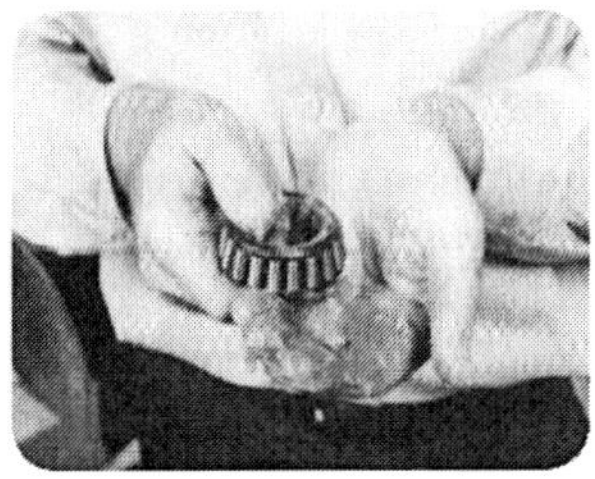

- ❖ Once the bearing is cleaned and dried, It is better to check the bearing for nicks, leftover dirt and damage.
- ❖ Properly see the bearings for wear and tear.
- ❖ Now dip the cleaned bearings in a protective lubricant or apply lubricant on the bearing using the brush.
- ❖ Rotate each bearing so that grease on it should properly spread.
- ❖ After applying lubricants, wrap them with waterproof paper and place in a clean box.

1.2.2 Lubricating 3-Point Hitch

After cleaning of tractor 3 point hitch, use proper grease, lubricants or engine oil to grease:

- ❖ lift link grease fittings
- ❖ ball joints
- ❖ draw bar

1.2.3 Lubricating Rear Axial

After cleaning of tractor rear axial, use proper grease, lubricants or engine oil to grease:

- ❖ rear axial bearing

1.2.4 Lubricating brakes and clutch pedals

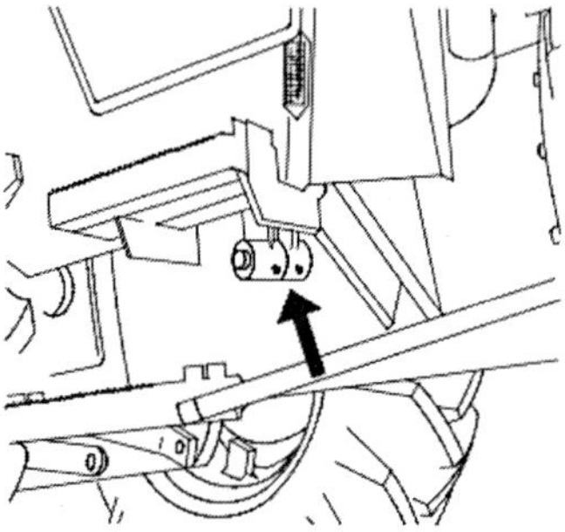

After cleaning of tractor brakes and pedals, use proper grease, lubricants or engine oil to grease:

- ❖ Nipple on left side of clutch pedal bearing
- ❖ Nipples on right side of brake pedal bearings

1.2.5 Lubrication of flywheel ring gear

After cleaning of tractor flywheel ring gear, use proper grease, lubricants or engine oil to grease by:

- ❖ removing the shield.
- ❖ remove the shield plug and apply grease to the ring gear

1.2.6 Lubricating Front Axle

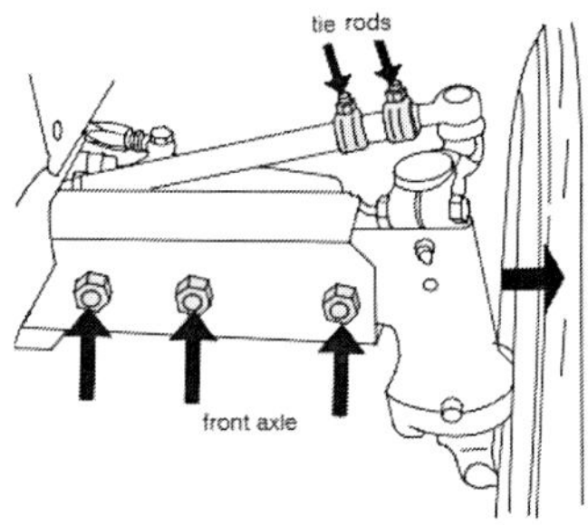

After cleaning of tractor front axle, use proper grease, lubricants or engine oil to grease:

- ❖ axle rods
- ❖ bolts of the axle
- ❖ nuts

1.2.7 Tractor parts that needs lubrication after certain working hours:

8 hours of running:

- ❖ Outer rotor bearings which are located inside belt housing, behind rotor drive pulley.
- ❖ Inner rotor bearings which are located on the center plate of the shredder.
- ❖ Standard PTO CV shaft having U-joints and extra fittings which is located under the bell housing.
- ❖ Center rotor bearings which are located between the rotors of each side of the shredder on the outside of the bearing mounting plate.
- ❖ Rotor to rotor coupler assemblies countersunk in rotor tube between 2 knife-carrier bars, approximately 4" from balancing ring on inner end of outside rotor.

25 hours of running:

- ❖ Telescoping PTO tube, which is located by lining up the hole in the PTO guard with zerk?
- ❖ Assist wheel pivot points which is located on each assist-wheel.
- ❖ Assist wheel caster swivel point located on each assist-wheel.
- ❖ Line shaft universal joints which is located on each side of the gearbox.
- ❖ Line shaft bearings which is located between the gearbox and belt compartments.
- ❖ Drive pulley shaft bearing which is located inside belt housing, behind drive pulley.

50 hours of running:

- ❖ PTO overrunning clutch, located near U-joint yoke
- ❖ Belt idler bracket, located inside the belt compartment.
- ❖ Front gauge wheel pivot point, one on each gauge wheel.

- ❖ Pull-type rear wheel lift tube.
- ❖ Assist wheel thrust plates above and on underside of each assist wheel.

100 hours of running:

- ❖ Wheel bearings
- ❖ Ratchet jacks.
- ❖ Gearbox.
- ❖ Wheel position locking pin tube, located on each assist wheel.

1.3 Follow basic operating instructions

Tractor operator has to refer to the tractor user manual and should follow the general working and safety instructions which includes:

- ❖ Gather information about the tractor before using it as insufficient knowledge about tractor driving and operation can result in accidents.
- ❖ Allow only specially trained and qualified person to operate the tractor.
- ❖ If the tractor is with a safety belt, then the operator should use it during the operation.
- ❖ Never allow a passenger to the cab unless there is an extra seat and hand-rail.
- ❖ Keep all warning plates clean. If the plates are damaged or loss, replace them by new ones.
- ❖ Prior to operation, carefully inspect the tractor and a trailed mechanism by starting.
- ❖ Trailed agricultural mechanisms and transport trailers should be fitted with rigid hitches excluding their swinging and rear-end collision with tractor during transportation.
- ❖ Prior to start an engine, use parking brake and set the gear shift lever to the neutral position while PTO lever in OFF position.
- ❖ Prior to starting the movement, warn the bystanders and the persons working on the trailed mechanisms with a horn.
- ❖ Never leave the moving tractor.
- ❖ Prior to leaving the cab, disengage the PTO, stop the engine, apply the parking brake, take off the starter switch key and set the battery disconnect switch to OFF position.
- ❖ Do not operate the tractor indoors without necessary ventilation as exhaust gases can lead in fatal outcome.
- ❖ In case of failure of the engine or steering control, stop the tractor immediately. In case of failure of steering control, a lamp of emergency oil pressure in HVSS should be lighted up.
- ❖ Do not work under lifted agricultural implements.
- ❖ If the front part of the tractor takes off when attaching heavy mechanisms and implements, in such case, mount the front weights.
- ❖ When operating the front lift, fill in the rear tires with liquid ballast.

- ❖ Prior to lifting and lowering the attached agricultural implements or when turning the tractor, make sure that there is no risk of hitting or snagging on somebody or any obstacle.
- ❖ When moving with attached mechanisms or implements, the operator should always apply a mechanism to lock on the attachment in lifted position.
- ❖ The drive shaft transferring the rotation from the tractor PTO to the driven components shall be enclosed.
- ❖ Make sure that any extra equipment/auxiliaries are installed correctly and that they are designed for application with the tractor.
- ❖ To avoid turnover, be careful when driving the tractor. Choose safe speed corresponding to road conditions, particularly when driving cross-country, passing the ditches, slopes and during sharp turns.
- ❖ When operating on the slopes, increase the tractor track to the maximum value.
- ❖ Do not make sharp turns under full load and high speed.
- ❖ When using the tractor for transportation works:
    - increase the tractor track to at least 1800 mm;
    - interlock the brake pedals, check and, if necessary, adjust the brakes to synchronous action;
    - check operation of the parking brake, brake pneumatic drive and trailer brakes;
    - check the state of the light and audio alarm devices;
    - transport trailers shall be fitted with rigid hitches and connected using safety chain or rope;
    - never drive downhill with switched-off gear and apply the same gear both down and uphill;
    - it is dangerous to work with the trailer without independent brakes, if its weight exceed half actual weight of the tractor.
    - disengage the FDA to avoid excessive wear and tear of the drive parts and tires;
    - do not use rear axle differential lock at the speed of above 10 km/h and during turns;
    - do not stop the tractor on the slopes, but if required, engage the 1st gear and apply the brake.
- ❖ when operating the equipment driven by PTO, make sure that PTO coupler shank gets fully stopped after stopping the engine before leaving the cab and disconnect the equipment.
- ❖ Do not wear loose clothes when operating PTO or near the rotating equipment.
- ❖ When operating stationary mechanisms driven by PTO, always apply the brake and lock the rear wheels from the front and behind.
- ❖ Make sure that the PTO coupler shank protection has been installed, if the PTO is not used, refit the cap of the PTO coupler shank.

- ❖ Do not clean, adjust or maintain the equipment driven with PTO, when the engine is running.

1.4 Procure maintenance tools that are certified for such operations from a recognized body.

Work for procurement of equipments will be done as per the following process:

- ❖ Pre-procurement
- ❖ Selection of Manufacturer
- ❖ Contract
- ❖ Completion

Selection of Manufacturer

The manufacture to be selected should fulfill following criterias:

- ❖ Experience in related products
- ❖ Certification from required authorities
- ❖ Required registration as applied
- ❖ Cost of equipment quoted
- ❖ Rates of bidders
- ❖ Selecting best and lowest bidder

Once the pre-procurement and manufacturer is selected for obtaining equipments, the work is divided among the purchaser and contractor. Following procedure should be considered while drafting the documents for procuring the implements or equipments.

Contract

The manufacturer should fill details of the product as mentioned in the procurement form:

TECHNICAL SPECIFICATIONS REQUIRED		TECHNICAL SPECIFICATIONS OFFERED PLEASE FILL OUT **ALL THE SPACES** BELOW TO SPECIFY THE EQUIPMENT BEING OFFERED FOR **THE OFFER TO BE VALID**	
Item	Description and purpose of the item	**Make/ brand and model of item**	
	A 4.0-5.0 tonne capacity, single axle, heavy duty, drop-side trailer for general agriculture purposes.		
Unit		**Name and address of the manufacturer of item**	
Total quantity of units required			
Country	**Project code**		

1.0.	**Technical Specifications:**	**General**	
1.1	Trailer carrying capacity:	4.0 – 5.0 tonnes	
1.2	Length (mm):	Approx. 4500	
1.3	Width (mm):	Approx. 2000	
1.4	Height (without extension) (mm):	400 - 550	
1.5	Platform level above ground (mm):	Max. 1100	
1.6	Chassis constructed of welded pressed steel channel.		
1.7	Heavy duty drawbar fitted with single towing eye		
1.8	Drawbar fitted with levelling screw jack		
1.9	Trailer axle:	Single axle, approx. 1500-70mm - 6 studs	
1.10	Fitted with manual parking brake.		
1.11	Tyres (with tubes):	11.5 - 15 12 PR (or similar)	
1.12	Trailer body: Floor constructed of 4mm (min.) sheet steel, all seams fully welded.		
1.13	Drop down and detachable trailer body sides constructed of 2mm (min.) pressed sheet steel for rigidity.		
1.14	Body fitted with a 2-part rear gate, height 1000mm.		
1.15	Body sides fitted with detachable extension panels constructed of 2 mm pressed sheet steel and attached to detachable corner posts Total height (side + extension) (mm):	1300	

Chapter 10

Check and Maintain Fluids Regularly/ Lubrication Points

2.1 Check the engine oil

- Park tractor on level surface
- Check the engine oil when engine is cool
- Pull out the dipstick, wipe it clean and dip in the oil sump.
- Ensure that oil level is near the upper mark between the upper and lower mark.
- If too low, add oil through the oil filler cap, but never excess 100 hours of service interval.

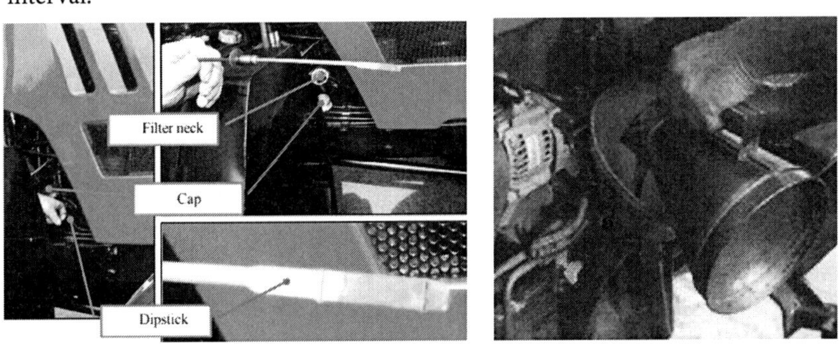

Fig Checking oil level and filling oil

2.2 Check the transmission fluid/ hydraulic oil

For checking the transmission fluid, follow the steps:

- With the oil is hot, remove the drain plug from the bottom of transmission housing and allow the oil to drain.
- Unscrew and remove the filler plug to allow faster drainage.
- Replace the drain plug.

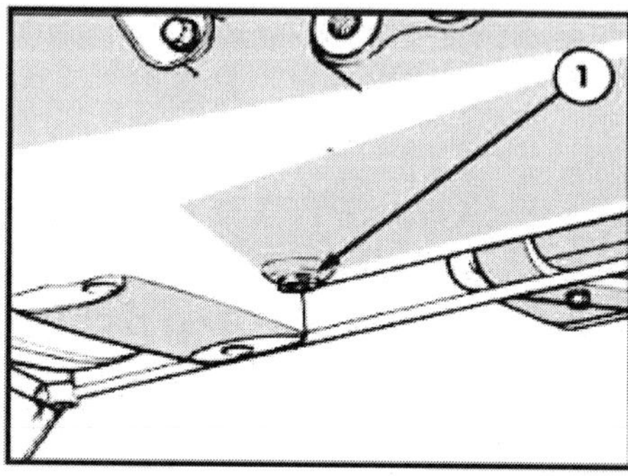

1.Drain plug

- Refill the transmission with clean oil.
- Pull out the dipstick and wipe it with a clean cloth.
- Replace the dipstick, pushing it fully in and check that the oil reaches the upper mark on the dipstick when it is removed from the transmission aperture once more.

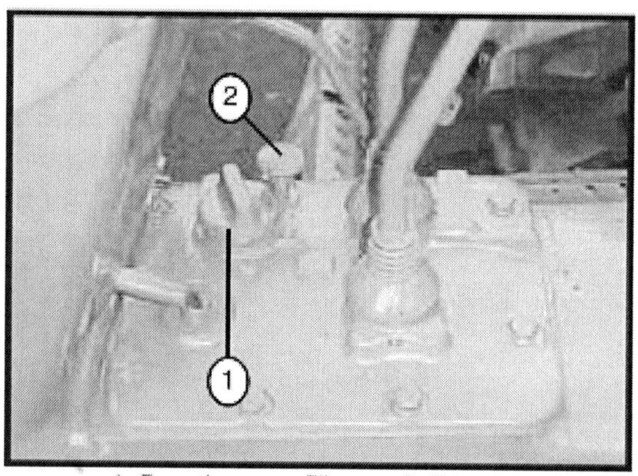

1. Breather cumfiller 2. Dipstick

- After refilling the transmission, start the engine to allow the oil to circulate throughout the transmission.
- After a minute, stop the engine and recheck the oil level by means of the dipstick.
- If required, top-up the oil to the upper mark on the dipstick.

2.3 Check the coolant in the radiator/ reservoir tank

When the tractor becomes heated up, in such case, tractor operator should check for the tractor radiator. Tractor operator can cool the tractor radiator by using the water. The operator should check that the water level of the radiator should be below from the top of the radiator.

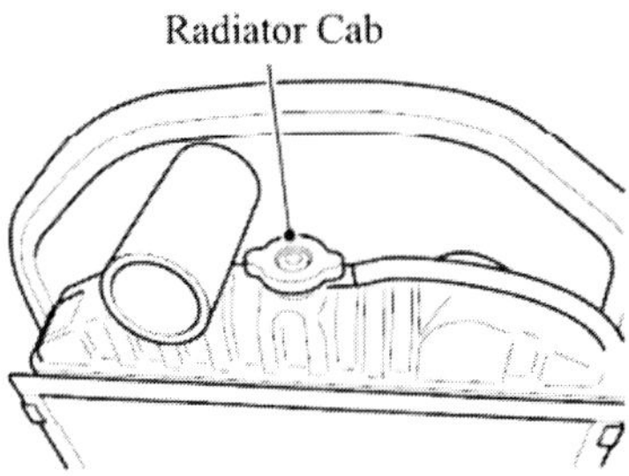

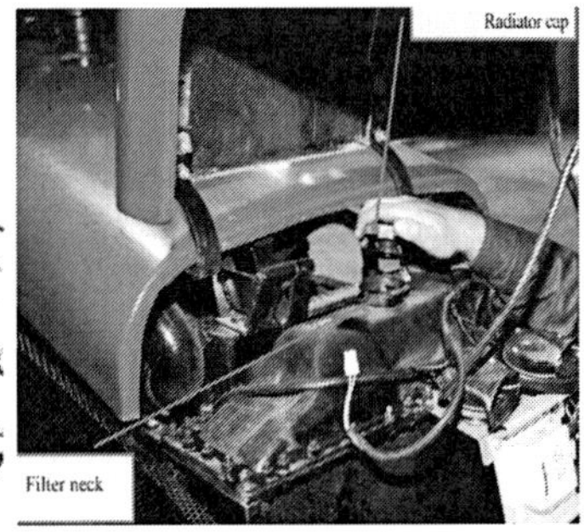

Once the radiator gets cooled, now tractor operator should check for the coolant level by removing the radiator cap.

Checking Coolant

- Park tractor on level surface

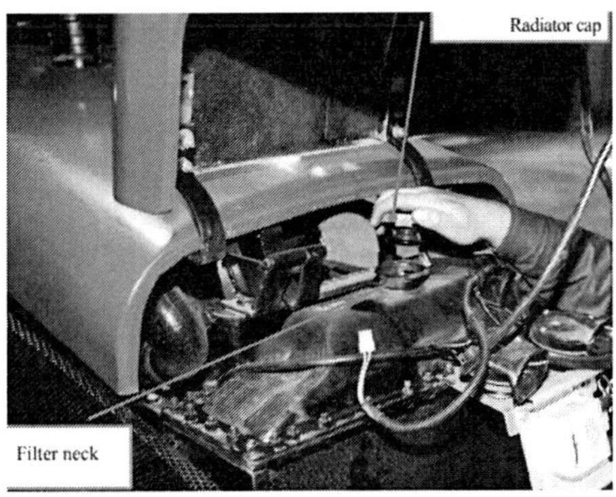

- Check the coolant when engine is cool enough.
- Open the hood.
- Remove the radiator cap.
- Ensure that the coolant is up to the filler neck and that it is clean with the correct anti-freeze or anti corrosion inhibitor in it.
- If low, add coolant.

Coolant should be levelled with upper end-face of filler neck and if not, add the coolant to the level.

2.4 Bleeding of fuel system

Tractor operator should do the following steps in order to bleed the fuel system which includes:

- Open of the bleed screw fitted on top of filter

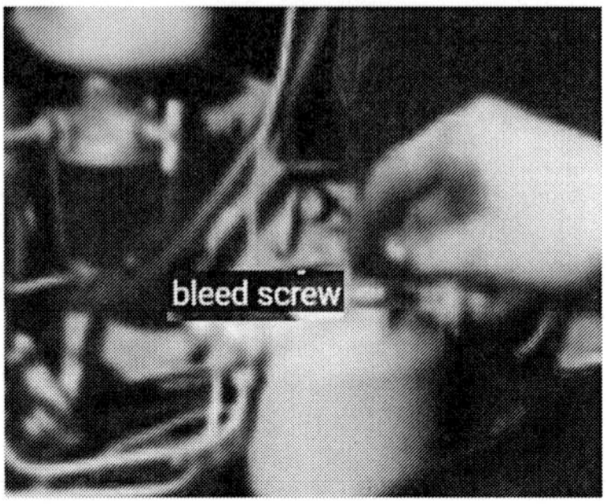

- With this, fluid will flow down from the tank and with the help of lift pump, we can circulate the fuel fill up the filter housing till the point where it comes out of this plug, there we have clean steady flow of fuel now.
- Refit the well
- Now fill the injection pump by removing the plate, we see it is completely empty.
- Fill the fuel by removing the screw

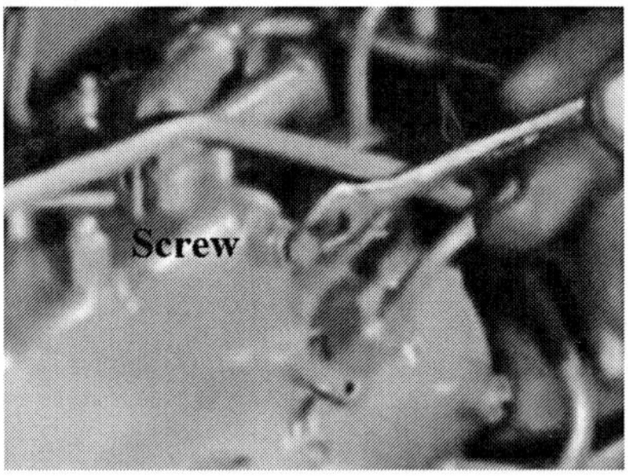

- Now pump the lift pump and you see that the fuel will flow out.

- After this, you will see that there will be no air bubbles.
- Refit the screw again and pump the lift boom for about 30 seconds which helps in circulating the fluid through leaking off system back to the tank.
- Slack off each injector pipes at the top.

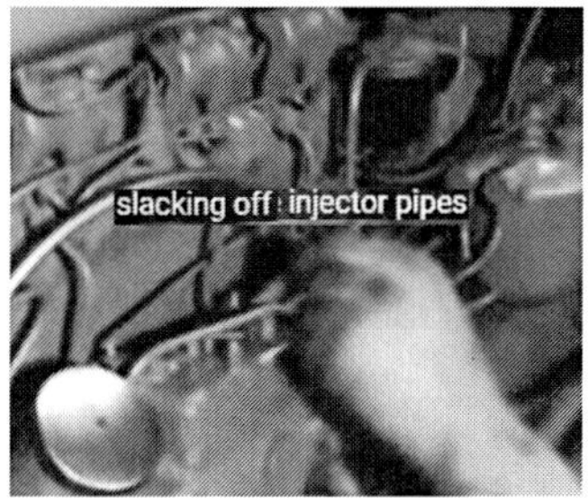

- Now turn the engine till fuel comes out of those so that we know that these are ready for bled off.

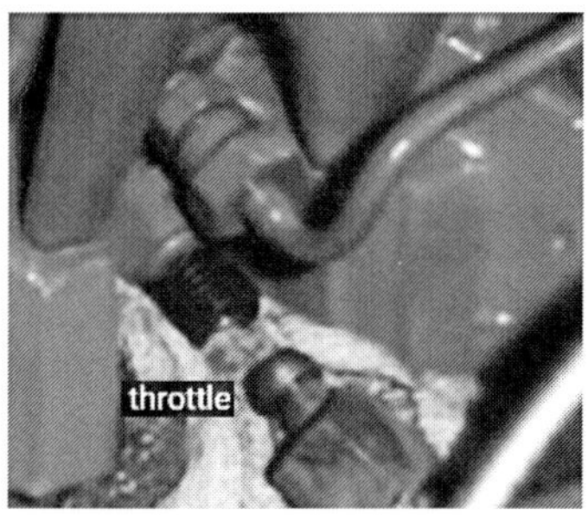

- Start the engine, so that injector pipes reaches to off and fit it again for full action of throttle

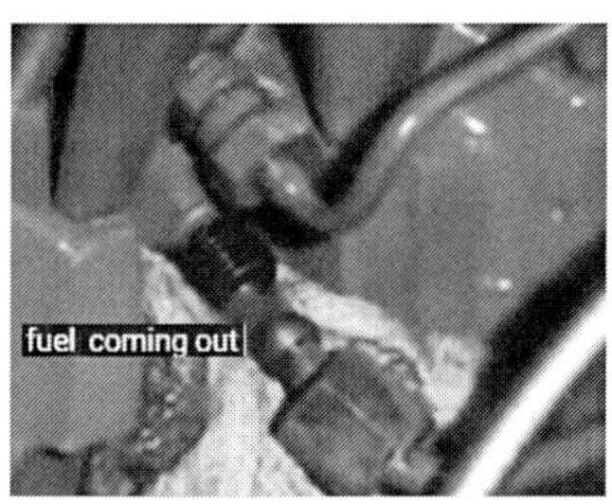

- Now with engine start watch he fuel coming out.

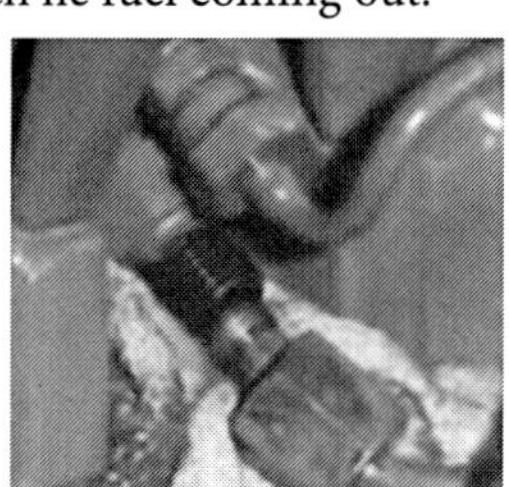

- Now fit the pipe back to the injector.

After going the above operations, the tractor is now ready to start. If you feel there is still air inside the system then in such case you have to repeat the above procedure.

2.5 Clean oil bath/dry type air cleaner

In the oil bath/dry type air cleaners, correct grade (weight) oil should be used as dirty oil will not be able to wash the particles down from the packing material. If the oil used is too heavy, it will not be carried up by the air on packing material or if the oil is too light, then also the oil is carried through the packing material and towards the engine. To keep the oil clean, the reservoir should also be cleaned and should filled with new oil each day or twice daily in case of dusty conditions.

Cleaning and Servicing of oil bath/dry type air cleaner

For cleaning the oil bath type air cleaner, follow the steps:

a) Dismantle the air cleaner assembly completely.

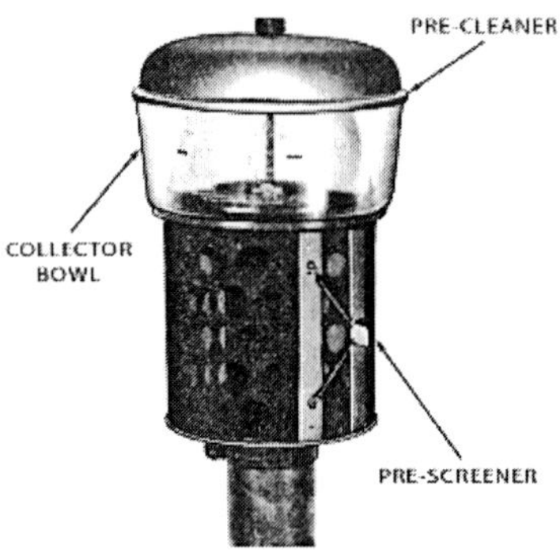

b) Remove the pre-cleaner and clean it with air under pressure by passing air in the opposite direction of its normal flow.

c) Remove the top screen and clean it with kerosene or diesel using a brush.

d) Remove the used oil, and clean it with diesel, and then refill with fresh mobile oil to the level as marked.

e) Clean the air hose and air pipe with compressed air.

f) Assemble the air cleaner assembly and attach it securely into its mounting located on the tractor.

For cleaning of dry type air cleaner, follow the steps:

a) Dismantle the dry type air cleaner assembly completely.

b) Remove pre-cleaner and clean it with air under pressure by passing air in the opposite direction of its normal flow.

c) Remove the top screen and clean it with kerosene or diesel with the help of brush.

d) Remove the used oil and clean the element by tapping or compressing of air or with detergent.

e) Clean air hose and air pipe with compressed air.

f) Assemble the air cleaner assembly and attach it securely into its mounting located on the tractor.

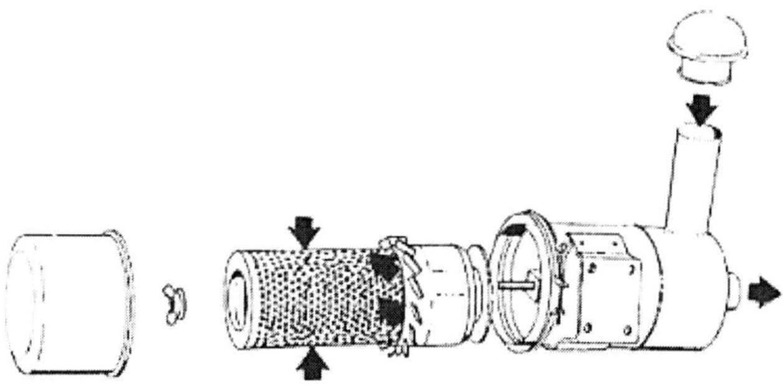

Fig dry type air cleaner

2.6 Check engine, transmission, hydraulic, steering oil level and topping up if necessary

2.6.1 Checking of Engine Oil level

By inserting oil dipstick in the engine and pulling it up will show the oil level as indicated on the stick. In the dip stick, high and low marks are present, which shows an exact oil level.

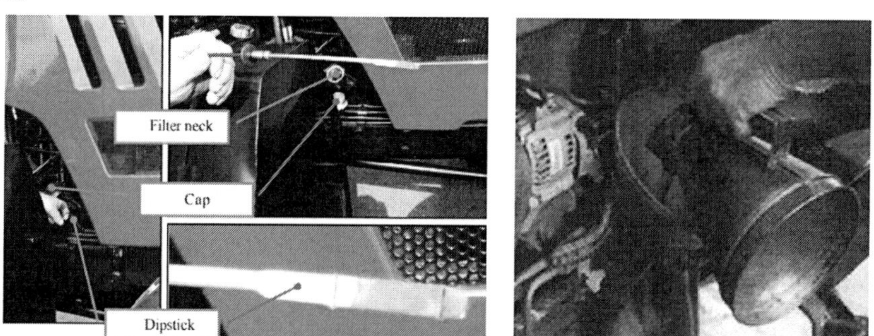

Fig Checking oil level and filling oil

2.6.2 Checking Transmission oil

- ❖ With the oil is hot, remove the drain plug from the bottom of transmission housing and allow the oil to drain.
- ❖ Unscrew and remove the filler plug to allow faster drainage.
- ❖ Replace the drain plug.

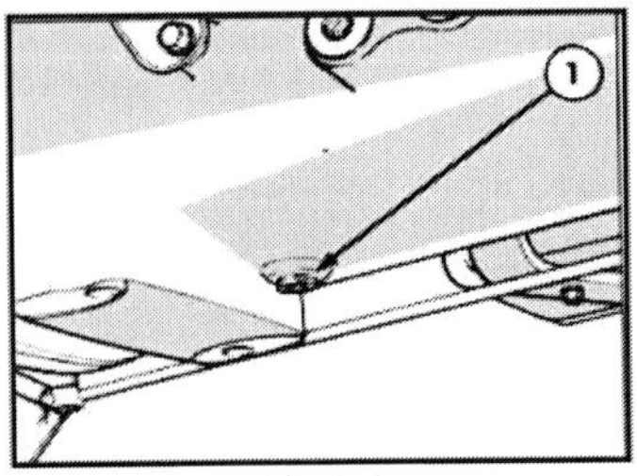

1.Drain plug

- ❖ Refill the transmission with clean oil.
- ❖ Pull out the dipstick and wipe it with a clean cloth.
- ❖ Replace the dipstick, pushing it fully in and check that the oil reaches the upper mark on the dipstick when it is removed from the transmission aperture once more.

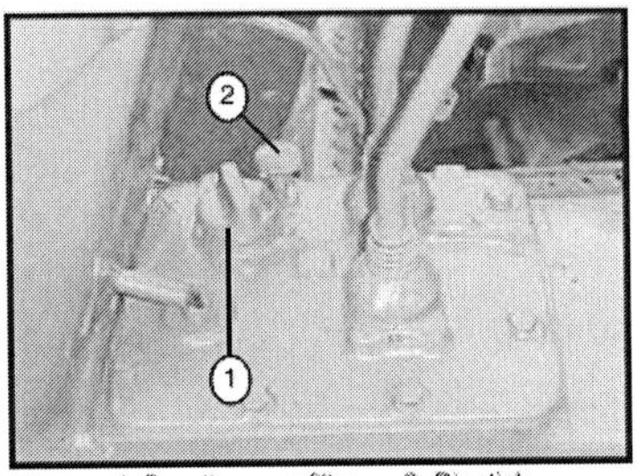

1. Breather cumfiller 2. Dipstick

- ❖ After refilling the transmission, start the engine to allow the oil to circulate throughout the transmission.
- ❖ After a minute, stop the engine and recheck the oil level by means of the dipstick.
- ❖ If required, top-up the oil to the upper mark on the dipstick.

2.6.3 Checking of oil level in steering

In tractors, steering system can be mechanical or with power. It is necessary to check the oil level in the steering system to avoid risk of losing control of the tractor. In modern tractors, steering fluid is part of hydraulic system which is drawn from same reservoir. In such case, when you change the hydraulic fluid, it also includes change of steering fluid.

a. Checking Fluid in Mechanical Steering System

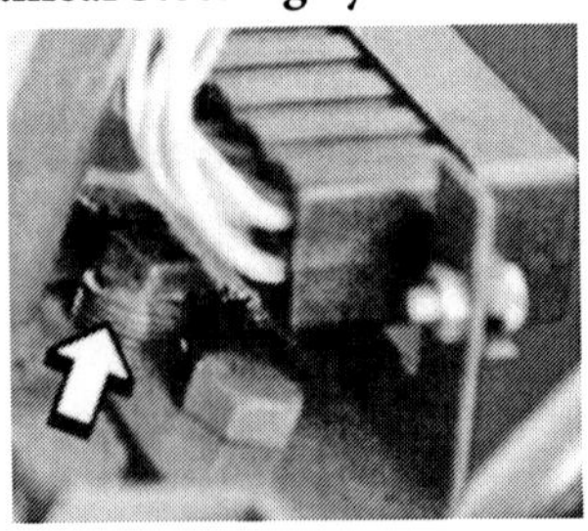

- To check the steering oil level, open the steering box.
- Unscrew the plug which is located on steering box.
- Check for the oil level.
- If the level of oil is low, add the oil.
- Tightly screw the plug.

b. Checking fluid level in Power Steering system

Power steering fluid is hydraulic oil which is used in high pressure hydraulic power steering systems and other vehicles. In many tractors, power steering symbol are present on power steering reservoir cap. If you are unable to find the power steering symbol, then check the owner's manual for power steering reservoir location, fluid level specification and fluid type.

If the fluid level is low, it will result in:

- Groaning noise
- Groaning noise on turning
- Increases steering effort

Steps

- Once the power steering reservoir is located, then initially check the area around it.
- Open the reservoir cap.
- In case the reservoir is wet, then fluid level can be high or results loose/damage cap seal.
- Check the fluid level from the marking or level shown.
- If the level is low, then top up the power steering fluid with funnel.

2.6.4 Check the hydraulic oil

All hydraulic systems start with oil. Oil is the lifeblood of the beast. Since oil is a liquid it cannot be compressed. That means it will flow and transmit power virtually anywhere. As with any fluid this form of power transmission needs to be channeled and told where to go. In simple terms that is what a hydraulic system does. The oil is the lubricant and vehicle for transmission of this power. Hydraulic oil is not just any oil. It is formulated to withstand a wide variety of temperatures and has additives that control its reaction to the rubber seals and plastic parts found in the components of the system.

- Hydraulic fluids perform several important tasks, including assisting the transmission, lubricating gears and hydraulic components, powering wet brake systems and maintaining consistent clutch operations.
- There is a wide variety of hydraulic fluids available on the market today, and not all created equally in terms of their performance or intended application.
- Many hydraulic oil are conventional, mineral-oil-based, zinc-containing type.
- High-performance tractor fluid is the best way for farmers to maximize years of reliable, trouble-free tractor life, increased productivity and profit.
- A wrong oil type can cause o-ring swelling and dissolve seals. Use only hydraulic oil recommended for your application.
- Hydraulic fluids should be changed as often as recommended in the owner's manual or more frequently if working in harsh or extreme environments.

2.7 Replace fuel filters and electrical fuses

2.7.1 Replace fuel filter

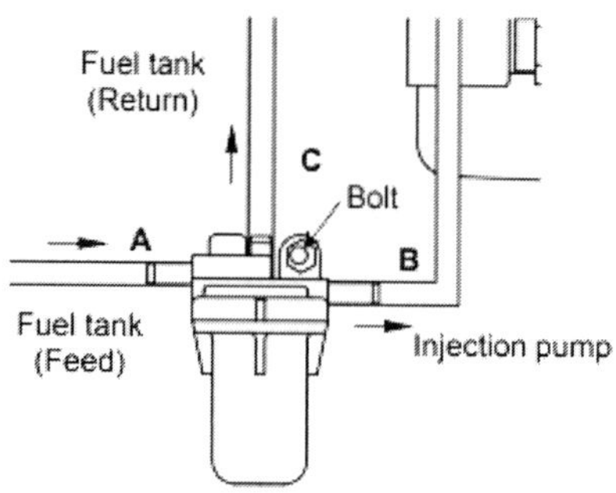

- For this, stop the engine and apply brakes.
- Adjust the container under the fuel filter.
- Prepare the plug to block the fuel hoses.
- Disconnect hose-A first, and plug the hose.
- Now disconnect hose-B and C using similar method.
- Remove the bolt from the filter.
- Replace the fuel filter with a new one.
- Connect the hose B, C and A step by step and tighten the bolt.
- Disconnect hose "C" for a seconds to bleed the air, and connect it again.
- After this, check the leakage around fuel filter.

2.7.2 Replace electrical fuses

- Loose wires makes off and on connections while damaged wire results in short circuits and can lead to reduce efficiency of components.

- If a fuse burns out again after it has been replaced, in such case, tractor mechanic/ operator should not replace the fuse with wire or a high capacity fuse, but should locate for the cause and try to rectify the fault, so that next time when the fuse is changed, it will not blow off.
- Where insulation is chafed or peeled off, recover the area with a good quality insulation tape.

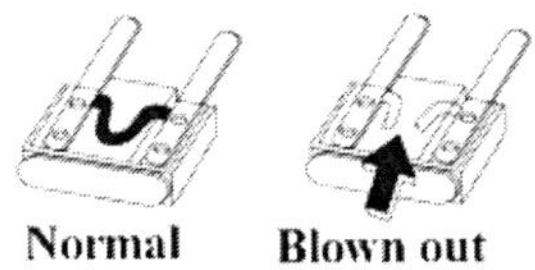

Main Fuses

- The wiring harness is equipped with 3 main fuses who's function is to preserve the wiring.
- However when a main fuse blows off, in such case, the entire circuit gets dead.
- Always check the reason & rectify before replacing the fuse of the same value.
- Blown off fuse gets discolored with fuse wire broken.

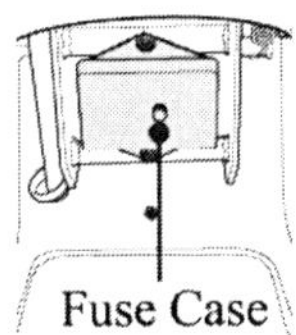

Replacing of Fuses

- The circuit has 8 blade type fuses in its wiring circuit
- When a fuse has blown off, replace it with other fuse of similar value.
- Using a large capacity fuse or wire burn out the wiring system.
- Use fuse tongs to replace fuses.

2.8 Check battery electrolyte level

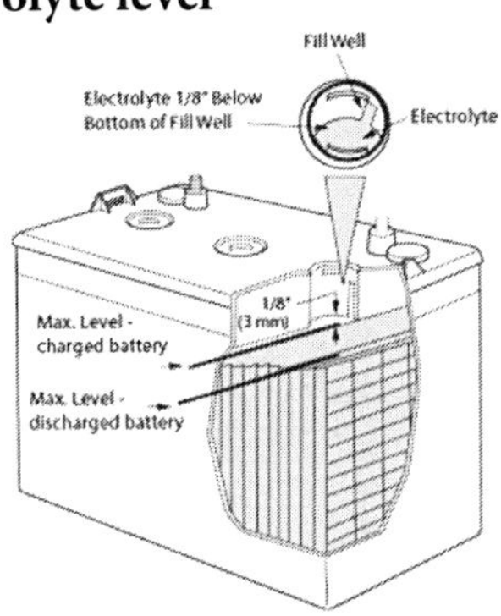

Electrolyte is a solution of acid and water and can cause severe burns. Battery in tractor is used for starting the tractor engine and to supply power to remaining engine parts. It comprises of cell compartments and is made of lead based plates which are submersed in solution of suphuric acid.

A hydrometer is an instrument which is used to test lead-acid batteries. As the density of battery acid changes, a hydrometer employs floats to check the density of the acid, thus indicating the charge.

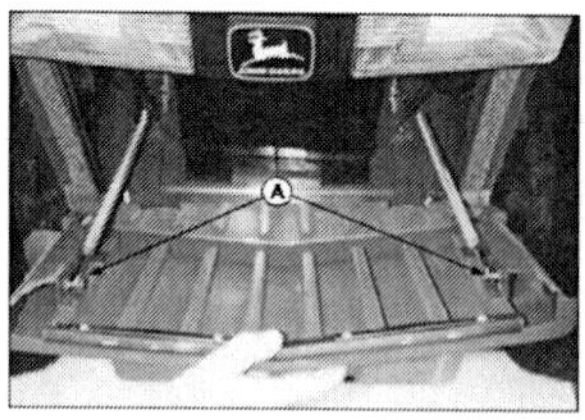

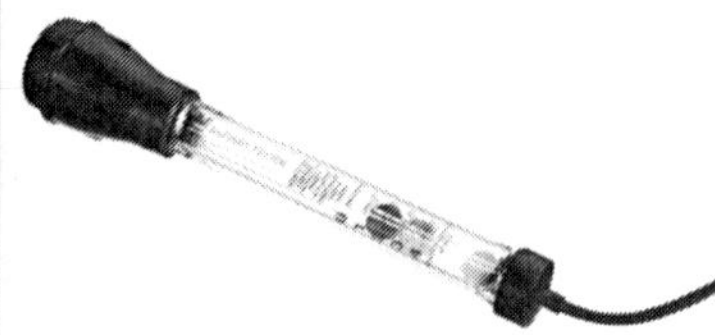

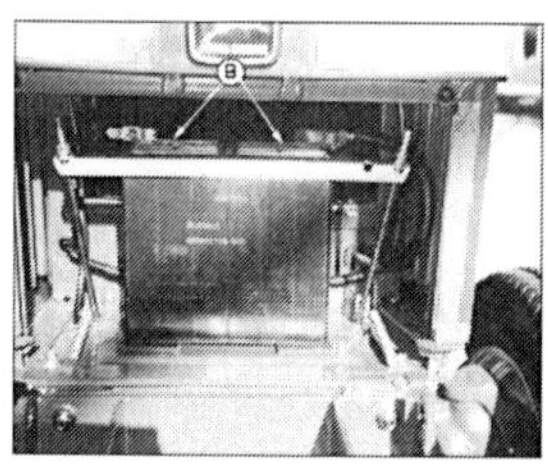

Before filling the battery with distilled water, check the electrolyte levels of the battery for battery water level.

To check the battery electrolyte level, follow the steps shown:

- ❖ Open the tractor locate the battery.
- ❖ Pull the grill out and disconnect the two springs (A).
- ❖ After locating the battery, remove the battery caps (B) to check the battery water.
- ❖ If the level of battery water is low as marked, then fill the cells inside the battery with electrolyte.
- ❖ Fill the electrolyte (D) of around 6 mm above plates (E).

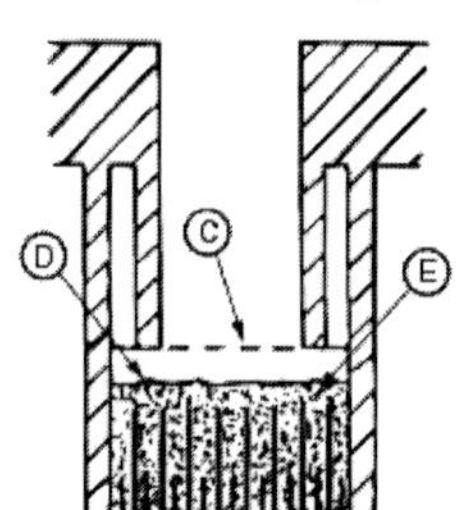

- ❖ You can also add distilled water.
- ❖ Close the cell openings of the battery with cell caps.

Chapter 11

Ensure Right Tyres Pressure

3.1 Ensure all the tyres have the right inflation pressure as per the usage of the tractor

3.1.1 Inflation pressure

Before moving the tractor, tractor operator should first check the tyre pressure. If the tyre pressure is less, then there are chances of tyre puncher and breakage. Check the tyre inflation with an instrument and keep optimum pressure as desired in the manual.

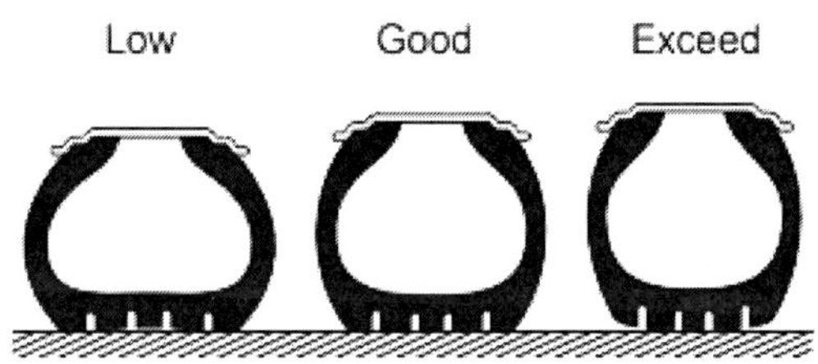

Inflation pressure in tyres should be checked on weekly basis. Recommended inflation pressures based on total load on tyres should be used. For accurate inflation, use a special low-pressure gauge with 10 kPa graduations. Gauges should be checked occasionally for accuracy. Always use sealing valve caps to prevent loss of air.

The optimum pressure for tractor tyres is determined by:

- ❖ tyre design,
- ❖ load on the tyre
- ❖ speed at which it work

Tyres under-inflated:

- ❖ Uneven wear and tear
- ❖ Shorten tyre life
- ❖ Damages sidewalls and tyres that slip on the rim
- ❖ 20% under pressure = 26% loss of performance

Tyres over-inflated:

- Increased soil compaction,
- Increased wear and tear
- Reduced operator comfort
- Shortened tyre life
- Increased fuel consumption

Good tire pressure will help in giving most out of tractor compactness. Improper tire pressure can, over time, lead the tractor to ruin. Tire pressure supports the structural integrity of a tire and will degrade the tire if it is not at the proper inflation levels. Uneven wear and poor traction are side effects of an underinflated tire which can lead to bead slip which occurs when the bead of the tire slips against the tire's rim creating more heat which finally destroys the tire.

Tire and tractor manufacturers agree that as little as a 20% over-inflation can reduce tractor performance by 30%. Proper tire pressure is something to monitor closely which can have a larger effect on pulling power, traction, ride quality and soil compaction.

Correct tire inflation reduces tire wear, increases fuel efficiency, and leads to less roadside breakdowns as a result of tire failures. Inflation recommendations are always based on cold air pressure, so it is recommend to check the tractor tire pressure before using it for the day. Tire pressure is affected by number of different factors from weather to usage.

3.1.2 Adjusting tyre pressure

- At the begining, inflate tyres to the lowest recommended pressure for the weight carried per tyre.
- If axle load changes, adjust the tyre inflation pressures as required.
- For determining load per tyre, divide the total axle weight by number of tyres per axle, taking into account extra weight on the rear axle due to mounted equipment.
- For optimum performance, set all tyres on a given axle to the same pressure.
- Observe traction outcomes and readjust accordingly.
- Check the inflation pressure often, preferably when the tyres are cold, such as in the morning.
- Gauge readings can be one to three psi higher when the tyres are warm.

3.2 Remove and refit tyres for puncture

Tractor operator needs to inspect the tractor tyres for correct pressure, puncture before staring for the work. Tyres with correct inflated and appropriate amount of tread are less likely to suffer from damage as compared to tyres which are worn or underinflated.

Many time the tractor operator see less pressure in the tyre and assumed a puncture. For such scenario, tyre should be inflated with correct air pressure and before use, should be standstill for 15 minutes, to check for actual puncture status. If there is a puncture in the tire, the air will tends to leak and the tyre gets settled at the base.

In case of puncture, the tractor operator should follow, the following procedure:

- ❖ Place the tractor on leveled surface.
- ❖ Place the jack at the axial to lift.
- ❖ Using wrench, open the nuts and bolt of the punctured tyre.
- ❖ Keep the puncture tyre on flat surface and manually take out the punctured object
- ❖ Release the air from the tyre to repair the puncher area.
- ❖ Use rim flange to quickly remove the tyre.
- ❖ Now, take the tube out of the punctured tyre and rub the punctured spot for applying puncture lotion.
- ❖ Using a piece of rubber tube piece and stick it with lotion on the punctured spot.
- ❖ Allow the area to dry.
- ❖ After 10 minutes, put the tube inside the tyre and adjust the air nozzle.
- ❖ Place the tyre on the rim and inflate it with required pressure.
- ❖ Screw all the nuts of the tyre.
- ❖ Release the jack pressure.

Chapter 12

Check Belts and Hoses

4.1 Hoses

- Hoses are used in many systems on every vehicle.
- They carry liquids such as fuel, oil, coolant and transmission fluid.
- They transport gasses like refrigerant used in the air conditioning system.
- They route engine vacuum to numerous parts, which may include the cruise control actuator, power brake booster, exhaust emission controls and electronic sensors for engine management system.
- The consequences of hose failure leak are directly tied to contents of the hose.

4.1.1 Symptoms of Hose failure

- The smell of gasoline when the engine is running
- Visible fluid leakage on the pavement under the tractor
- A sweet burning smell
- A smoky burning smell
- Visible smoke or steam from underhood
- Glowing of 'Check Engine' light on the tractor dashboard
- High engine temperature gauge reading or an illuminated engine temperature warning light.

4.1.2 Hose Inspection

Cold Engine Inspection:

- To avoid accidental injury or burns, inspect for small leaks and check hose condition when the engine is cold.
- Check for hoses that are cracked, frayed or missing pieces.
- Look for hoses that have been cut or abraded from contact with other components.
- Inspect around hose ends for dampness, deposits, or buildup of dried coolant which shows minor "cold leak", which takes place when tractor warmed up and cooled down.

- A hose that feels hard or makes a “crunching” sound when squeezed is deteriorating and should be replaced.
- Hoses that are extremely soft, sticky or oil-soaked are bound to fail and should be replaced.

Warm Engine Inspection:

- Use extreme caution working around hot engine parts.
- Drive the tractor long enough for the engine to reach full operating temperature.
- Park the tractor and shut off the engine and then open the hood.
- Inspect the upper radiator hose.
- If upper radiator hose is collapsed, in such case, the radiator cap may be faulty.
- Look carefully at all the hoses to spot any swollen areas that would indicate internal damage or weak spots.

4.1.3 Checking of Hoses

It is the responsibility of tractor operator to check for different hoses in the tractor after every 200 hours of operation or 6 months, whichever comes early.

Radiator Hose

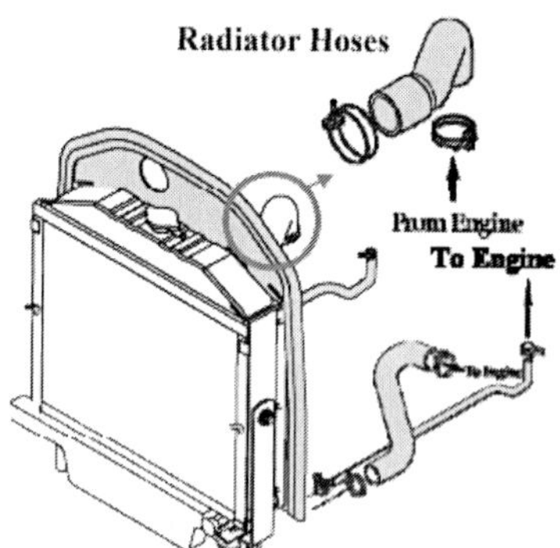

Radiator hoses are import part of the cooling system that carry hot coolant from the engine through the radiator and circulate it back to the engine. There are two hoses, the upper and lower, which are thick and sturdy to handle both high pressures and immense heat.

For checking the radiator hose:

- Park tractor on level surface
- Shutoff engine. Allow engine to cool enough.
- Open the hood.
- Check the radiator hoses for damage and leakage.
- Check the radiator hoses for damage and leakage.

- ❖ Replace if required.
- ❖ Tighten Clamps
- ❖ Check for any leakage around the hoses and hose clamp.

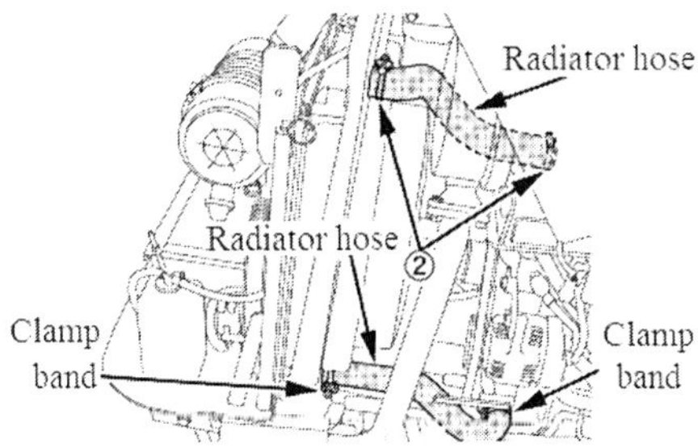

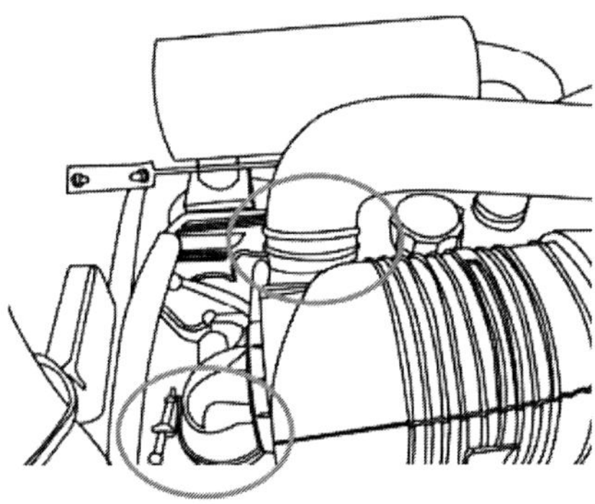

Air Intake Hoses

For checking the air intake hose:

- ❖ Park tractor on level surface
- ❖ Shutoff engine. Allow engine to cool enough.
- ❖ Open the hood.
- ❖ Check the air intake hoses for damage.
- ❖ Tight the clamps

Hydraulic hose

Check of Hydraulic hoses

- ❖ Switch tractor off – wipe all hose and fitting surfaces with a clean rag.
- ❖ Wear eye protection, gloves and close fitting clothing.
- ❖ Restart tractor and cycle all the hydraulics until the oil reaches operating temperature.
- ❖ Lower attachments to the ground then turn engine off and check that all cleaned areas show no signs of dampness.
- ❖ DO NOT place hands around hoses or connections when the system is under pressure.

4.2 Belts

When a belt fails and breaks, the components it was powering will get immediately cease to function which lead to something very serious such as loss power steering, coolant circulation or engine operation.

4.2.1 Types of Belts

To meet the various needs, automobiles uses different types of belts:

- Serpentine Belts
- Timing Belts
- V-Belts
- Fan belt
- Drive belts

4.2.2 Symptoms of Belt failure

When the belt fails, the tractor operator should experience the following symptoms:

- A squealing noise as the engine is accelerated
- A squealing noise when the steering wheel is turned fully in either direction when parking
- A slow, rhythmic slapping sound at idle
- A whirring or grinding noise from a pulley when the engine is running.

4.2.3 Belt Inspection

Belt used in tractor will result in failure if they are damaged, broken or loose. To spot these signs, tractor operator should perform a visual belt inspection and belt tension check once a month.

For inspection of belt:

- ❖ Carefully inspect the belts by looking at their edges and undersides for signs of any wear.
- ❖ Check for the glaziness of the belt which appears when belt area which is in contact with the pulleys becomes slick and shiny which results due to lose and slips on the belts.
- ❖ Checking for any belt cracks as cracked in V belts should always be replaced immediately.
- ❖ Check for the belt which gets oil soaked due to leakage from engine gaskets and other underhood sources, as such belts used to slip and become glazed.

Belt with above symptoms should be replaced and tractor operator should rest sure about the repair of leaks that could further lead to such problems.

4.2.4 Tension in belt

When tension of belt exceeds the standard value, then loose the bolt and adjust the tension.

For adjusting belt tension:

- ❖ Loosen two bolts of alternator and hinge bolt of adjusting plate. Insert the bar between the crank case and pull the bar to apply tension to the belt.
- ❖ With proper tension, tighten the bolts attached to the adjusting plate of alternator.
- ❖ If the belt tension becomes normal, tighten the other bolts and check the tightening torque.

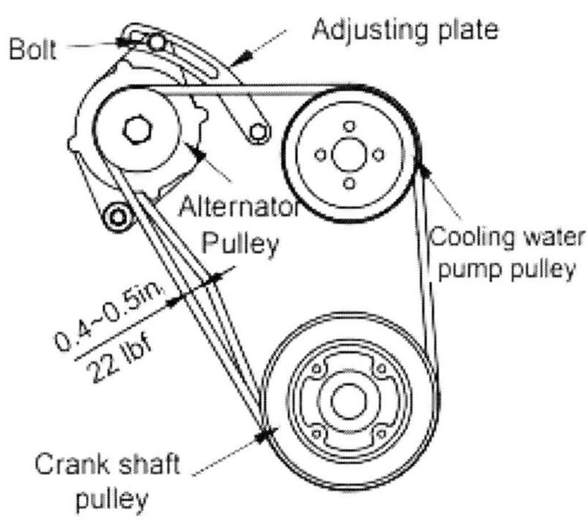

4.2.5 Fan belt play

Fan belt in tractor when loose or tight will result in overheating of an engine and may result in damage of an alternator. For this reason, it is necessary for a user to check regularly, the tension in tractor fan-belt.

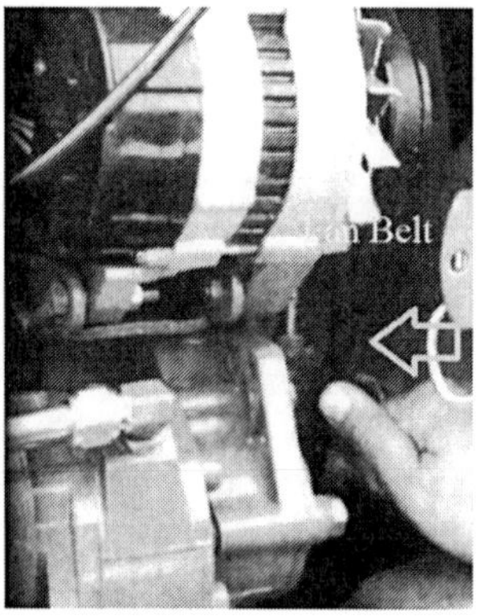

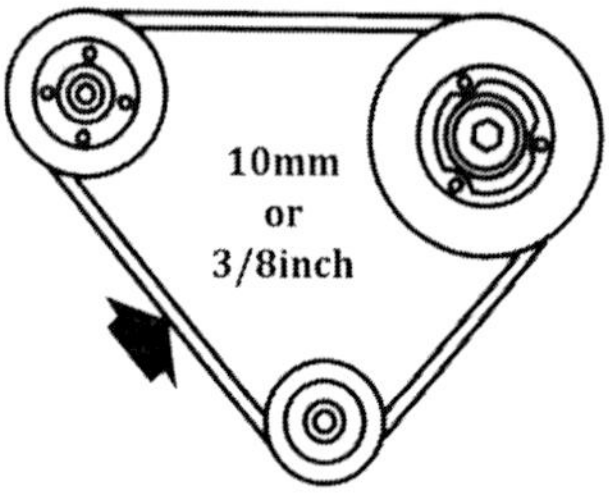

Check

To check the tension, press the longer portion of the belt between the alternator and crankshaft pulley with certain force and then loosen the belt to around 10mm or ⅜ of an inch.

Adjustment

If the fan belt is loose, then follow the steps to get it tightened:

In case of improper tension in the belt, pivot the alternator bolt. The other end of the alternator which is attached to the slider, allows the alternator to change its position, so that the belt can be tighten or loosen.

Now loose the alternator pivot bolt and the bolt which goes through adjustment strap. By loosening the bolts, the alternator will be loosen and can be in moving position.

Now add tension to the drive belt and insert pry bar above the alternator pulley. Gently apply pressure upwards so that it adds tension to the drive belt. After moving the belt in desired tension, tighten the adjustment bolt to lock the belt position. Finally tight the adjustment bolt and check the belt tension again. If the tension has decreased/increased, then again adjust the belt by following the similar process of adjustment.

When the belt is positioned correctly, then tight the pivot bolt in other side of the alternator. Finally check the belt tightness by pivoting all the bolts and verify belt deflection at the longest point.

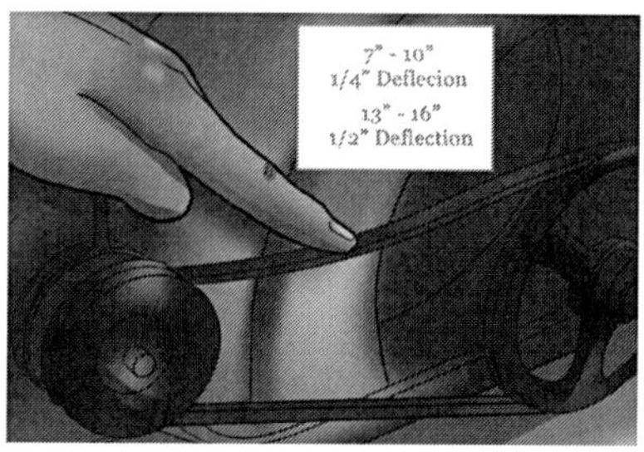

4.3 If fittings or connections are leaking, tighten them or replace the seals.

Belt and hoses in the tractor are attached with the help of seal, clamp and adhesives. When the engine gets hot after running for certain hours, in such case there are chances for any leakage or loosening of parts.

Tractor operator should check for any leakages or loosening of hoses or belt parts:

❖ Checking to see hoses and hose clamps are tight and not damaged.

❖ If hoses and clamps are found worn or damaged, replace or repair them at once.

4.3.1 Tighten of loose parts

While checking the belt and hoses, if the tractor operator locate any loose part, he/she should be able to tight the loose parts. If the part is loose, it will either vibrate, gives humming sound or can be broken down. There are many types of nuts, bolts and washers which are used to assemble the fan, radiator or pulleys, so when such parts get loosened, they will affect the other parts and there are chances of severe damage or hazards.

4.3.2 Checking leakage in radiator hose

- For checking the leakages in radiator hose, turn off the engine.
- Clean the leakage area with cloth.
- Restart the engine so that the radiator hose involves in its functions.
- Inspect for the leakage such as drips or trickles of fluid/oil.
- If the leak is still not visible, again clean the suspected areas.
- Check for leakages while running an engine.

Chapter 13

Brake and Clutch Adjustment

5.1 Ensure proper adjustment of the brake and clutch, and make sure the brakes and clutches free play are adjusted properly

Tractor operator after entering into repair and maintenance of clutch and brakes finally will have to make themselves assured that the proper adjustment, alignment and free play of clutch and brakes are performed as per user manual with correct specification as desired. For this, he/she has to manually operate the clutch and brakes for checking of items as desired.

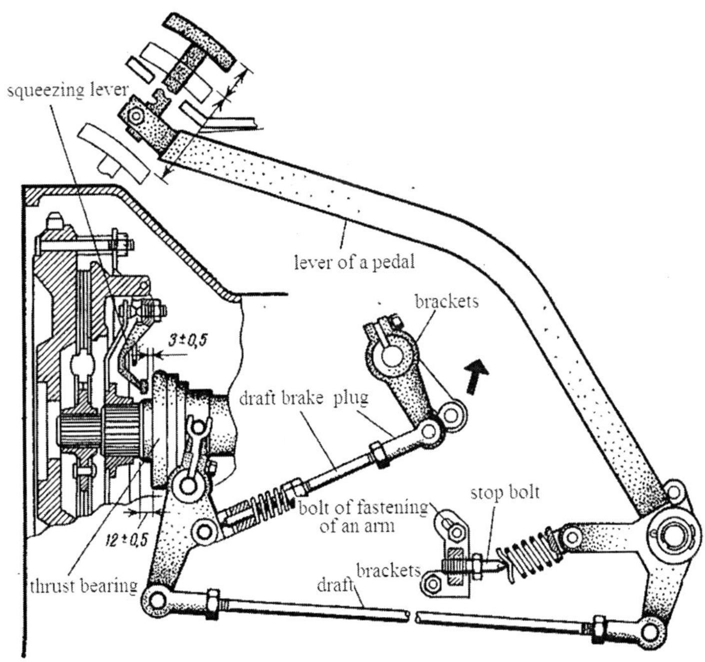

5.1.1 Adjustment of clutch

Tractor mechanic/operator should adjust the clutch, by using the following instructions:

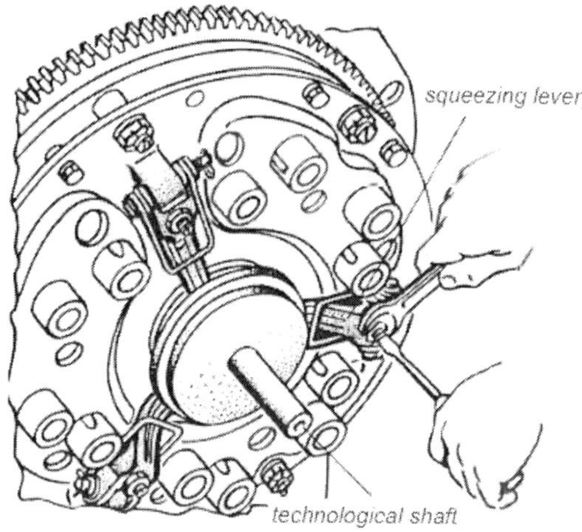

- The free travel of the clutch pedal is controlled by the rotation of the linkage.
- In the released position, the pedal lever should rest against the cockpit polish.
- If lever is not resting against the cockpit polish, then slowly remove the stop bolt from the bracket.
- Also, loose the bracket mounting bolt and turn the bracket clockwise towards the spring.
- If the clutch release mechanism is correctly adjusted and the dimension is 12 ± 0.5 mm, then the gap between the thrust bearing and protrusions of the levers will result as 3 ± 0.5 mm.

5.1.2 Adjustment of brake

Tractor mechanic/operator should adjust the brake, by using the following instructions:

- Disconnect the draft from an arm
- Turn the bracket to the right anticlockwise until it stops
- Rotate the fork, which increases the length of the linkage till the bracket and plug fit freely.
- Rotate the plug till the total length of the rod is reduced by 7 mm
- With this position, connect the plug to the bracket
- Once the adjustment is done, screw the plug with the nut

5.1.3 Adjustment of free play

- Free-play in brakes is a measurement of how far the brake pedal moves before the master cylinder initially begins to apply.
- It shows the amount of pedal movement before the pushrod touches the piston inside the master cylinder, before the piston starts moving.

- ❖ It is critical on all braking systems in order to keep the Vent Port open in master cylinder which on closing will increases brake fluid pressure due to heating up of brakes.
- ❖ It can be measured using a ruler or tape measure.
- ❖ Too small free play is not good while too much is rarely a problem due to pedal reserve

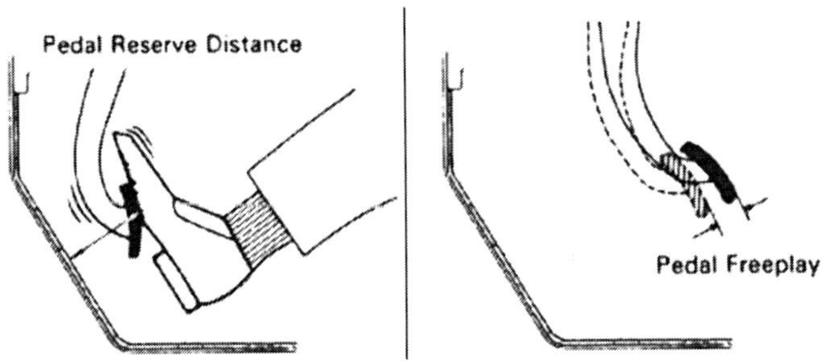

Adjustment:

- ❖ Adjust the length of push rod by loosen 15mm lock nut on push rod with open-end wrench.
- ❖ Turn the nut on the push rod till it feels like flushing up against piston in master cylinder.
- ❖ Turn the rod back about half a turn and check the free play at top of brake pedal.
- ❖ Repeat till free play as measured at top of brake pedal within 5-7 mm
- ❖ Now tight the lock nut and check again

5.2 Ensure that brake paddle latch should be engaged while driving on road.

- ❖ Engage the brakes by pushing down the brake pedal.
- ❖ Pushing the pedal down harder applies more air pressure.
- ❖ Letting brake pedal, reduces the air pressure and releases the brakes.
- ❖ Releasing the brakes allows compressed air to go out of the system, so air pressure in tanks is reduced.
- ❖ Test the brakes by driving at low speed.

Chapter 14

Check working of all Electrical Systems and Gauges

6.1 Ensure right temperature, oil pressure and hour meter

6.1.1 Gauges

Gauge has coil of oval-section tube which joins the exterior capillary tube, so when water temperature increases pressure, it causes the oval-section tube to move in outward direction and so pointer needle moves around the dial. There are gauges and meters in many tractors which comprises of:

- fuel pressure gauge,
- oil pressure gauge,
- water temperature gauge,
- hour meter,
- hydraulic pressure gauge
- temperature gauge

1 Temperature gauge:

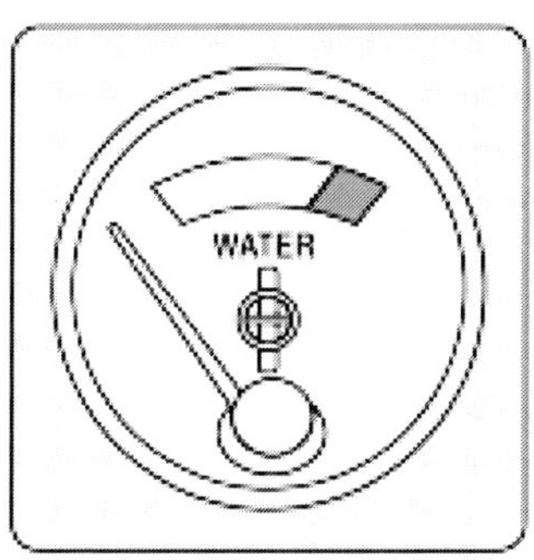

- In this gauge, the needle of gauge instrument will show engine coolant temperature.
- In this gauge, there are light and dark zones.

- Needle in light zone shows that engine temperature is OK.
- Needle in dark zone, shows that tractor engine is overheated.
- For checking the temperature, the engine should be shut off and allow to cool down so as to determine the source of problem.

It is observed that very hot weather, also produce a high gauge reading, but in such case, gauge will return to normal under cooler conditions. After a cold start, pointer should indicate normal operating temperature within 10 minutes.

High reading in temperature gauge could show faults related to:

- fan belt broken or slipping;
- electric drive failed;
- radiator blockage;
- radiator hose collapsed;
- cylinder block furred;
- shortage of water in system;
- failure of water pump or thermostat;
- failure to remove radiator muff or raise blind

Check right temperature is maintained in the gauge

In tractor, most important temperature to know is about cooling-system water where calibration is no more than cold rising to hot, with the normal operating temperature (80 to 85°C). It is very important that the correct engine operating temperature should be maintained at all times as low operating temperatures contribute to formations of destructive acids and harmful deposits.

2 Check for right oil pressure

When tractor engine initially started, oil pressure shows normal on the gauge, but as the engine gets warms up, the oil pressure drops to the low range. This happens, as oil on warming up becomes thinner which results in leakage and flows more easily. It leads to non functioning of oil pump which due to it becomes worn and will not able to maintain the oil pressure.

Also, low oil pressure results from bearings which has oil all its ways that lead to wornness making the oil to flow more easily due to warmness making oil to flow around engine morc casily/quickly as comparcd to pump.

For checking the oil pressure, user has to check for:

❖ oil pressure gauge, as low oil pressure reading is not taken by the gauge correctly.

❖ condition of oil, as it can turn into jelly which block the oil strainer and oil filter.

❖ correct and clean oil filter, as oil filter is often too small to give full pressure at higher RPM's when filter is blocked or too small.

❖ oil pump, as damage or worn pump will not cavitate oil in the pump.

❖ oil gallery plug, as due to looseness and damage will not result in correct oil pickup and sucks air.

3 Hour meter

Hour meter is a gauge/instrument which tracks and records the elapsed time shown in hours and tenths of hours. It is used to log the running time of equipment to assure proper maintenance of expensive machines or systems which can be in terms of replacing, changing or checking parts, belts, filters, oil, lubrication or running condition in engines, motors, blowers, fans etc.

The reason for using an hour meter is for maintenance. If you rely on the actual running time of an engine to schedule maintenance from the last service, rather than calendar days or miles driven, there is less guesswork and greater correlation to the true operating life of the equipment.

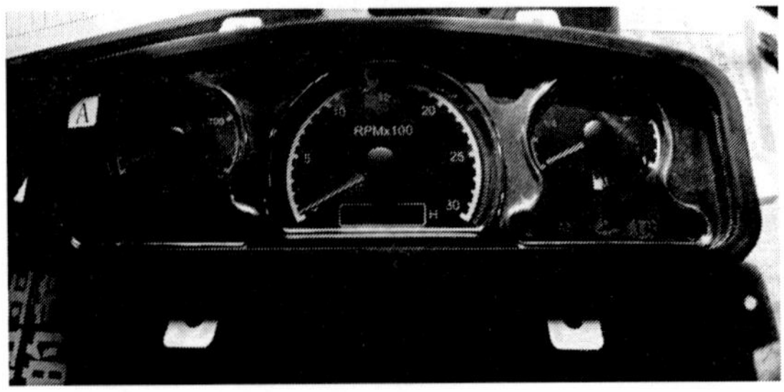

On the dashboard where all meters are installed, see for small rectangular 8-digit odometer. Newer tractors have digital hour meter that is displayed on the dashboard. Read the hours on the hours odometer having 6 digits, a decimal point and then 2 digits for a partial hour reading.

Check the adjustment of hour meter

Hour meter is a gauge or an instrument which keeps the track and records of elapsed time, that usually shows in hours and tenths of hours. These are used to log running time of equipment to assure proper maintenance of machines or systems.

In this, the maintenance involves replacing, changing or checking parts, belts, filters, oil, lubrication or running condition in engines, motors, blowers, and fans, to name a few. It is used for:

- tracking the running time which exists between scheduled service.
- monitoring and tracking the elapsed time/hours
- recording actual running hours

6.2 Electrical system of tractor

The electrical system of the tractor is classified as:

- Starting system
- Lighting system

Starting System

Starting system comprises of starter which is a DC motor that gets heavy current from the battery. It is used for intermittent service under heavy overload. In this, the four poles are in series connected and give high starting torque. When the circuit is completed by the solenoid switch, the pinion of the motor automatically meshes with the crankshaft ring gear. The reduction ratio between these two is 8:1 or 16:1. As the engine is started the pinion is thrown back to its initial position due to inertia effect.

Lighting System

- In the lighting system of tractor the lamps or bulbs are used for warming purpose and lights for illumination purpose.
- The main lamps include the headlamps, the tail lamp, the number plate lamp, the tail lamps the direction indicator lamps and dashboard lamps.
- The complete lighting circuit has number of individual circuits for a single lamp or a pair of lamps each with its own switch, live connection and earth connection.
- The accessories like horn are connected parallel across the battery terminals having a switch.
- The dashboard gauges like oil pressure gauge, the temperature gauge do not have separate switch which gets operated as soon as ignition switch is on.

6.2.1 Ensure working of headlights, brake lights and horns.

a. Headlights

The light switch, which forms a combination switch with the turn signal light switch, has two positions OFF and ON. Current passes through the light circuit as shown.

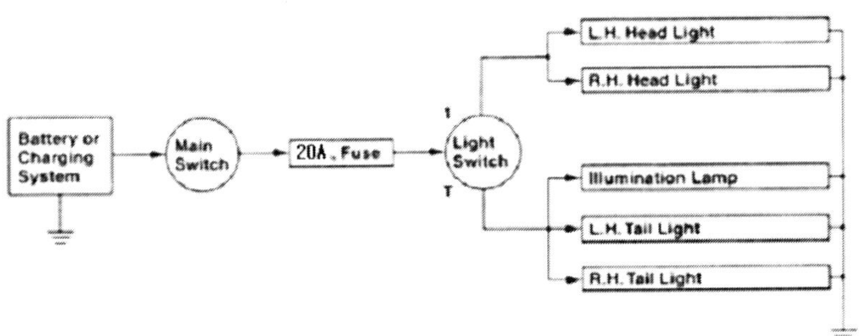

- While checking the headlight of the tractor, make sure that:
- Centers of the headlight low beams should land on the road about 30 meters ahead of the tractor,
- The left angle of the light should be straight ahead while the right angled should ahead slightly toward the edge of the road.
- Headlight high beams should aim just below the horizon.

Adjusting the headlights

- Before adjusting the headlight, check that the tires are inflated properly and that the tractor is carrying usual load.
- Park the tractor on level surface such that the headlights should be 25 centimeters from a wall or garage.
- Turn on the low beams and draw a cross on the wall to mark the center of each beam.
- Now back the tractor perfectly straight 7.5 meters.
- Each headlight has two adjusting screws, one at the top or bottom and one at the side to adjust the aim, tighten or loosen the screws.
- Now set each headlight at low beam to 5 centimeters below the horizontal lines with left even with vertical line while right 5 centimeters to right of vertical line.
- On four light system, adjust the high beams in same way as low beams.
- On two light system, high beams are automatically adjusted with low beams.

b. Checking and Adjusting Roof light:

- In vertical direction, a point of illuminated area in plane of road towards left from longitudinal vertical plane passing through the headlight centre should not be more than 30m from the front outline of the tractor.

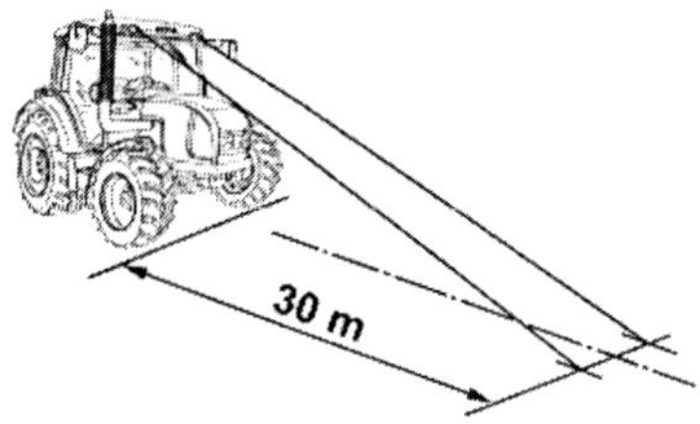

- In horizontal direction the headlight beams shall be parallel with the longitudinal axis of symmetry of the tractor.
- Carry out checks of adjustment of the headlamps at un-laden weight of the tractor.
- The front roof headlights may be used during traffic on ground roads only if there is a front-carried mechanism is coupled to the tractor or an equipment covering the main headlights.

c. Brake lights

- Brake lights depend on tractor's electrical system which runs on series of switches and fuses to protect circuits.
- If one or more of tractor brake lights isn't working properly, in such case:
  - brake light system fuse is blown
  - brake light bulbs are burned out
  - brake light wiring switch is broken

6.2 2. Checking of Horn

In tractor, horn button is present on the centre of steering wheel for an operator to use. The button has a wires which are attached to the main horn which is present in the chassis.

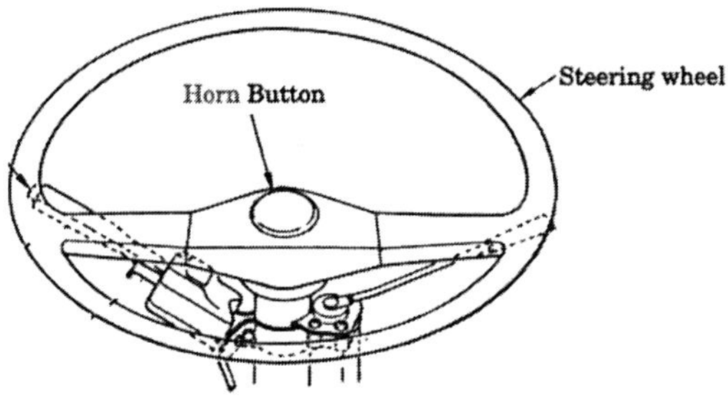

Horn should be checked before leaving for the day work as it is essentially required to reduce any hazard on field or on the roads.

To check the horn, press the horn button. If the horn is not working, the tractor operator should check for the following:

- ❖ Remove the horn button cover to see the loose connection of wires.
- ❖ Check for the wires which are going through the button knob till the actual horn.
- ❖ See for any wire looseness or breakage of wire at the actual horn side.
- ❖ Check the battery, as the horn wires are attached to the battery.

6.3 Avoid Overloading

Overloading has been recognized to be both a safety concern as well as a cost concern.

6.3.1 Risk of overloading

Overloading of tractor will pose the following risks:

- ❖ The tractor will be less stable, difficult to steer and take longer to stop.
- ❖ Tractors react differently when the maximum weights which they are designed to carry are exceeded.

- Overloaded tractor can cause the tyres to overheat and wear rapidly which increases the chance of premature, dangerous and expensive failure or blow-outs.
- The driver's control and operating space in the overloaded vehicle are diminished, escalating the chances of an accident.
- The overloaded tractor cannot accelerate as normal, making it difficult to overtake
- At night, the headlights of an overloaded tractor will tilt up, blinding oncoming drivers to possible debris or obstructions on the roadway
- Brakes have to work harder due to riding of brakes and since the tractor is heavier due to overloading.
- By overloading the tractor, tractor operator has to incur higher maintenance costs to the tractors in terms of tyres, brakes, shock absorbers and higher fuel consumption.

6.3.2 Recommendations

It is highly recommended that the tractor operator should read the user manual before overloading of tractor which can be done by following ways:

- Know the weight of the tractor – both in terms of permitted axle weight and gross vehicle weight
- The gross tractor weight is the maximum permitted weight along with any extra load, if there.
- Distribute the load appropriately to avoid overloading of axles
- Tractor weights should be checked using a weighbridge

Chapter 15

Attachment and Detachment of Implements

7.1 Tractor implements

Box Blade

- It is a type of implement which is for leveling and smoothing of uneven ground.
- It is pulled behind the tractor, so it is able to leave a pristine, untouched surface.

Mower/Cutter

- It is used for clearing grass and light brush.
- It is pulled behind the tractor and many models connect to the power take-off (PTO) to turn the cutter blades.
- Some implement are powered by their own independent motors.
- Some implement are better suited for handling refined landscaping work while others are better suited for knocking down bigger brush in rougher territory.

Loader

- It is a tool with many potential uses.
- It is a set of arms that attach to the front of the tractor which allow it to lift large loads into the air.
- It comes with a large bucket in the front which can be filled with dirt, cut brush, fertilizer, etc.
- It is extremely helpful on farms where it can help moving large quantities of manure, mulch, topsoil and more.

Post Hole Digger

- It is essentially a large auger attachment which works with tractors.
- It is extremely helpful in case of large amount of fence line to maintain.
- It is helpful for landscapers who regularly install large quantities of new plants.
- With this, digging plenty of tree holes is much easier.

Rotary Tiller

- ❖ It uses rotating tines which looks like a large auger to turn up topsoil.
- ❖ It is helpful for turning a cover crop into the ground or for preparing the soil for planting.
- ❖ Once the ground is used by the implement, it will have several inches of open, soft soil which becomes ready to accept seeds.

Spreader

- ❖ It is designed to fling a product which could be seeds or fertilizer in wide, even layer around a field.
- ❖ It uses rotating head and centrifugal force to shoot the product to a particular distance.

7.2 Using attachments

- ❖ Before using any attachments or implements, always read, understand and follow the manufacturer's safety manual.

- ❖ Only use equipment that is in good repair and make sure that all guards or shields are in place and operational. Guards around the power take-off shafts, gear box and other rotating/spinning equipment are very important.
- ❖ Check that all hydraulic lines are in good condition, securely connected and leak-free.
- ❖ Add front counterweights to tractors with rear-mounted attachments for improved steering control and stability.
- ❖ Use correct brackets and recommended sizes and types of bolts for mounting equipment.
- ❖ Use jacks, hoists or blocks to position equipment.
- ❖ Avoid crushing and entanglement points and do not stand between a tractor and an implement.
- ❖ Ensure that safety locks and catches are fastened.
- ❖ Attempt to line up the hitch holes from the tractor seat, then shut off the engine before dismounting to attach the hitch.
- ❖ Do not crawl under anything supported only by a hydraulic lift. Always use blocks,

jack stands, or other suitable and stable supports.

- ❖ Make sure all guards are in place and repair or replace the damaged guards and shields.

7.3 Attachment and Detachment of implements

- ❖ Tractor implements are interchangeable everyday which can be attached and detached to the tractor.
- ❖ The attachment of implement is attached to the tractor on an attachment hitch system which is located on the rear of tractor which gets pulled behind the tractor during operation.
- ❖ Larger tractors provide for attachment of implements to the front of the tractor using a front attachment hitch system that is similar in concept to the rear attachment hitch system.
- ❖ Such front attachment hitch systems are not commonly employed on tractors that are used in some farm or agricultural operations.
- ❖ Implements are attached to an attachment hitch system that is located in a midsection of the tractor beneath a tractor frame.

7.3.1 Attachment of Implement

A) Hitching

Tractor operator during the hitching of an implement should:

a) Make sure that the correct hitch system is used.

b) Use controls from the operating position.

c) Not stand between the tractor and other machines or behind them.

d) Not stand with feet under on or near drawbars.

e) Ensure that jacks, skids and other supports are used and maintained.

f) Communicate clearly in case of any help.

B) Drawbar

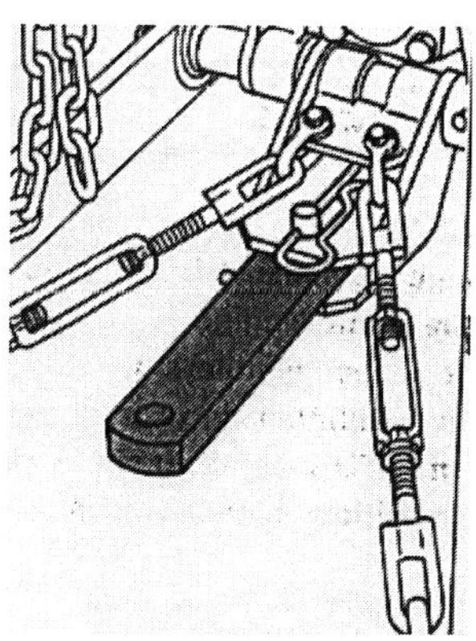

The drawbar is used for towing implements such as trailers. The tractor operator should:

a) Only pull from the drawbar.

b) Pull away slowly.

c) Keep brakes in tandem while towing.

d) Use proper drawbar pins with locking pins to connect the tractor and implements.

Correct

Incorrect

e) Remember to plug in other related attachments if relevant. These could be electrical, hydraulic or brake connections. If used, ensure the pipes and cables are clear of the ground and will not snag drawbars or three-point linkage

f) Use the drawbar correct configuration for type of implement or trailer being towed as failure to do, will lead drawbar pin to bend or break when travelling over undulating ground

Pick-up hitch

Many modern tractors are fitted with pick-up hitches. These enable heavy trailers to be attached without any manual lifting or the use of jacks and without leaving the operators seat. The system operates from the rear linkage of the tractor and has a mechanical lock, to secure the hitch without putting excessive loads on the hydraulic system.

1. Hitch pin or hook
2. Locking support hooks
3. Hitch

C) Three point hitch

Tractor operator should clearly follow the given attachment procedure which includes:

i. Backing of tractor up to the implement, and aligning the lower link ball ends with the implement hitch pins and using safe stop procedure.

ii. Pushing the hitch pin through the ball and secure it with the linch-pin.

iii. Attaching of right hand lower link to the implement, using the levelling lever to adjust the height of the link if necessary.

iv. Attaching of top link to the implement frame and to the top link attachment point on the tractor.

v. Leveling of implement from side to side and front to back.

vi. Adjust the check chains or stabilisers to minimise side swing on the implement.

7.3.2 Detachment of implements

Tractor operator should follow the procedure while detaching an implement which is attached to the tractor using:

i. Selecting of level area, as this will make detaching much easier.

ii. Lowering the implement, and applying the brakes.

iii. Using of safe STOP procedure.

iv. Dismounting from the tractor and disconnecting implement end of the top link.

v. Detaching the lower links and replacing the linch-pins to avoid losing them

Index